SPACEX STARSHIP ENGINE CORE RETURNS TO BASE BOOK

How Groundbreaking Tech is Making Space More Accessible

MARCUS T. HOOKS

COPYRIGHT

Copyright©2024 Marcus T. Hooks. All rights reserved. No part of this publication may be reproduced, distributed, or transmitted in any form or by any means, including photocopying, recording, or other electronic or mechanical methods, without the prior written permission of the publisher, except in the case of brief quotations embodied in critical reviews and certain other non-commercial uses permitted by copyright law

TABLE OF CONTENTS

COPYRIGHT 1
TABLE OF CONTENTS 2
INTRODUCTION 4
   SpaceX Starship Engine Core Returns to Base 4
CHAPTER 1 12
   The Genesis of Starship and Super Heavy 12
CHAPTER 2 22
   Reusability in Space Travel: Why It Matters 22
CHAPTER 3 35
   The Science of Controlled Landings 35
CHAPTER 4 51
   Testing and Trials: The Journey to Success 51
CHAPTER 5 64
   Chopsticks: The Game-Changing Landing Mechanism 64
CHAPTER 6 77
   From Earth to Mars: Starship's Interplanetary Mission 77
CHAPTER 7 93
   Environmental and Regulatory Challenges 93
CHAPTER 8 108
   The Future of Space Travel: Commercial and Scientific Impacts 108
CHAPTER 9 126
   What's Next for Starship? 126

CONCLUSION 141
  The Vision and Reality of SpaceX's Starship System 141

INTRODUCTION

SpaceX Starship Engine Core Returns to Base

The year is 2024, and the world stands at the precipice of a new era in space exploration. As humanity gazes toward the stars, one name consistently drives the vision of an interplanetary future—SpaceX. With its groundbreaking advancements in reusable rocket technology, SpaceX has redefined the boundaries of space travel, making what once seemed impossible, possible. Among these achievements, the return of the *Starship Engine Core*—a pivotal component of the Super Heavy booster—marks a critical milestone not just for SpaceX, but for the entire aerospace industry.

This introduction will explore the significance of this engineering feat, delve into the technological innovations that made it possible, and consider the broader implications for space exploration, scientific research, and humanity's future as a spacefaring civilization. It sets the stage for a deeper understanding of how SpaceX has revolutionised not only rocket science but also the global aspirations to reach beyond Earth.

A Leap Toward Reusability

The concept of reusability in rocket technology is not new, but the execution of a fully reusable system has remained elusive until now. Before the advent of SpaceX's innovations, rockets were traditionally built as single-use vehicles. Launching into space was an expensive endeavour, requiring the construction of a new rocket for every mission. This limitation was a major barrier to frequent space travel, exploration, and the larger vision of interplanetary colonization.

SpaceX's *Starship Engine Core* returns are a testament to the company's vision for making space travel affordable and accessible. The idea of returning key rocket components like the engine core to the launch tower—ready for rapid reuse—was once a dream, but now it is a reality. This breakthrough not only reduces the cost of launches but also accelerates the frequency with which rockets can be sent into space. This technological leap represents a monumental shift in the aerospace industry.

At the heart of this transformation is SpaceX's drive to develop a fully reusable system. Much like how commercial airplanes revolutionised air travel by being able to fly repeatedly with only minor maintenance in between flights,

SpaceX's reusable rockets aim to do the same for space. By recapturing and reusing major components of the rocket, such as the *Starship Engine Core*, SpaceX is pushing the boundaries of how quickly and affordably humans can access space. Elon Musk, the visionary CEO of SpaceX, has long maintained that reusability is the key to unlocking the future of space travel.

Technological Mastery: The Chopsticks and Mechazilla

The return of the *Starship Engine Core* to the launch tower was made possible by SpaceX's inventive system known as "Chopsticks" and the *Mechazilla* tower. These mechanical arms are designed to catch the descending rocket components after launch, a feat that requires pinpoint precision, robust engineering, and cutting-edge technology. The visual spectacle of the massive *Starship* booster being gently lowered by these "chopsticks" is as awe-inspiring as it is technically complex.

The precision required for such an operation is unprecedented in aerospace engineering. The rocket must be guided down at precisely the right speed and trajectory to align with the tower's arms, ensuring a safe and controlled return. The *Chopsticks* system eliminates the need for traditional rocket landings, which require vast amounts of fuel and carry significant risks of failure. Instead, by catching the booster mid-air, SpaceX bypasses these challenges, further enhancing the rocket's reusability and reducing turnaround times between missions.

This innovation is an engineering marvel in itself. Imagine the immense weight of a fully fuelled rocket descending from the heights of space, moving at breakneck speeds. The system must not only withstand the tremendous forces but also act with surgical precision to ensure a seamless catch. SpaceX has effectively redefined what is possible with large-scale rocket components.

The *Mechazilla* tower, which features these robotic arms, is a monumental achievement in design and functionality. It represents the epitome of modern aerospace ingenuity, combining structural strength, mechanical precision, and state-of-the-art computer algorithms. This system opens new doors for SpaceX's vision of rapid reusability and represents a giant leap toward the company's goals of reducing the cost and time required for space missions.

From Earth to Mars: The Bigger Picture

While the return of the *Starship Engine Core* is an impressive technological accomplishment, it is only one part of a much larger plan. SpaceX's broader ambition extends far beyond Earth's orbit. The development of reusable rockets is critical to Elon Musk's grander vision: establishing human colonies on the Moon and Mars. The reduction of launch costs through reusable technology will make these long-term, interplanetary missions not just a possibility, but a practical reality.

SpaceX has already secured a role in NASA's Artemis program, which aims to return humans to the Moon by 2026. The *Starship* vehicle, with its fully reusable Super Heavy booster and *Starship Engine Core*, has been selected as the lunar lander for the Artemis missions. But SpaceX's ambitions stretch even further, with plans to establish a human settlement on Mars. The success of the *Starship Engine Core* returns and the operational efficiency it brings are stepping stones to this ultimate goal.

It is no longer science fiction to imagine humans living on Mars within our lifetimes. With reusable technology at the forefront, SpaceX has shortened the timeline to interplanetary travel. The significance of this accomplishment cannot be overstated—by making space travel more frequent, reliable, and affordable, SpaceX is transforming the prospects of human life beyond Earth.

Environmental and Economic Implications

The benefits of reusable rockets extend beyond space exploration. One of the key advantages of SpaceX's reusable *Starship* system is its potential environmental impact. Traditional space missions generate significant waste, with spent rocket stages and fuel being left to decay in the atmosphere or remain as space debris. By returning and reusing rocket components like the *Engine Core*, SpaceX reduces this environmental footprint.

In addition, the economic implications are profound. Space travel has long been associated with astronomical costs, limiting participation to only the wealthiest nations and corporations. Reusability changes this equation. With every successful return of the *Starship Engine Core*, SpaceX reduces the overall cost of launching payloads into space. This opens new opportunities for commercial ventures, scientific research, and even space tourism. As costs decrease, space becomes a more accessible frontier for businesses, universities, and entrepreneurs alike.

Furthermore, the reusability of rocket components creates new job opportunities in engineering, manufacturing, and related industries. SpaceX's success is driving growth not only in the aerospace sector but across the broader economy.

The ripple effects of reusable technology are already being felt as more players enter the commercial space race, spurred on by the cost-effective model pioneered by SpaceX.

The Future of Space Exploration

The return of the *Starship Engine Core* is not just a triumph for SpaceX but a victory for humanity's future in space. It demonstrates that we are closer than ever to making space travel routine, efficient, and sustainable. With each successful test flight and return, SpaceX is refining the technology that will one day take us beyond Earth to the Moon, Mars, and possibly further.

In the coming decades, the innovations sparked by SpaceX's reusable technology will likely become the standard for all space missions, both governmental and private. The age of disposable rockets is quickly coming to an end, replaced by a future where rockets can be launched, returned, and reused in a matter of hours or days. This paradigm shift will have far-reaching effects on everything from satellite launches to deep-space exploration missions.

In conclusion, the return of the *Starship Engine Core* to its base is not just a technological achievement—it is a symbolic moment in human history. It marks the beginning of a new era in which the dream of regular, affordable space travel is no longer a distant aspiration, but an emerging reality. As SpaceX continues to push the boundaries of what is possible, the world watches in anticipation, knowing that each new breakthrough brings us one step closer to the stars.

CHAPTER 1

The Genesis of Starship and Super Heavy

The origins of SpaceX's *Starship* and the *Super Heavy* booster stem from a bold and ambitious vision: to make space travel affordable, sustainable, and ultimately interplanetary. Elon Musk, the founder of SpaceX, has long been fascinated with the potential of human exploration beyond Earth. His dream is not simply to send astronauts on brief trips to space but to create a future where humanity can live and thrive on other planets. In particular, Musk's goal is to establish a colony on Mars, which he sees as crucial to the survival of our species in the long term. SpaceX, therefore, was built on the foundation of reducing the costs of space exploration, making it scalable, and developing technology that could be reused.

The creation of the *Starship* and the *Super Heavy* booster represent the most tangible manifestation of Musk's vision. These groundbreaking spacecraft are designed not just for missions to the International Space Station (ISS) or lunar exploration but for deeper, more sustained missions into the solar system. The origins of these spacecraft lie in Musk's recognition that the existing technology for space travel, with its single-use rockets and expensive missions, was insufficient for his broader aspirations.

Early Concepts and Inspirations

The idea of reusable rockets was not new when Musk conceived of the *Starship* and *Super Heavy*. NASA had attempted reusability with the Space Shuttle program, but while the Shuttle itself was reusable, the boosters and external fuel tanks were largely single-use, and refurbishment costs were extraordinarily high. The Shuttle's lifespan was cut short in part due to these costs, and its overall mission architecture fell short of delivering cost-effective space travel.

Musk recognized that for humanity to become a truly spacefaring species, the technology needed to change radically. His solution was to build rockets that could be fully and rapidly reusable, reducing the cost per flight and allowing for frequent space missions. Early concepts at SpaceX, such as the *Falcon 1* and *Falcon 9*, were steps in this direction. These rockets were designed to be partially reusable, with the first stage capable of returning to Earth for refurbishment and reuse. This was a major leap from

13

traditional expendable rockets, which are discarded after a single use, adding millions to the cost of each mission.

The *Falcon 9* was SpaceX's first major success in reusability, and it paved the way for the development of the *Starship* and *Super Heavy*. The *Falcon 9*'s first stage was successfully landed after multiple missions, proving that rockets could be reused. While the technology was still in its infancy, it laid the groundwork for larger, more ambitious projects.

Evolution of the Starship and Super Heavy System

The journey from the early concepts of *Falcon* rockets to the development of *Starship* and *Super Heavy* was marked by significant technological advancements and a relentless pursuit of innovation. Musk's vision required not just a reusable rocket but one that could carry enormous payloads to the Moon, Mars, and beyond. This need for power and scalability led to the development of the *Super Heavy* booster, which would serve as the backbone for the *Starship* system.

SpaceX engineers designed *Super Heavy* to be the most powerful rocket booster ever built. With 33 Raptor engines, it generates an unprecedented amount of thrust, making it capable of lifting the *Starship* spacecraft and its cargo out of

Earth's atmosphere. The *Starship* itself is also powered by multiple Raptor engines, which are designed to be not only powerful but also highly efficient, capable of using different types of fuel depending on the mission requirements.

The key innovation in *Super Heavy* and *Starship* is their reusability. Unlike the *Falcon 9*, which can only partially return to Earth, both the *Super Heavy* booster and the *Starship* spacecraft are designed to be fully reusable. The *Super Heavy* booster is equipped with landing legs and grid fins, allowing it to return to the launch site after deploying the *Starship* into space. The *Starship* itself is designed to land on the Moon, Mars, or other celestial bodies and then return to Earth, making it the first fully reusable spacecraft in history.

Overcoming Challenges

The development of the *Starship* and *Super Heavy* has not been without its challenges. Early prototypes of the *Starship*, such as *Starhopper* and *SN-series* models, suffered from failures and setbacks. Explosions, structural failures, and other issues plagued the early stages of development, but Musk and the SpaceX team persisted. Their approach has always been to "fail fast and learn faster," using each failure as an opportunity to improve the design.

One of the biggest challenges in the development of *Super Heavy* and *Starship* has been the sheer scale of the project. The size and power of the *Super Heavy* booster present unique challenges for engineers, particularly when it comes to re-entry and landing. The booster must descend through Earth's atmosphere at high speeds, and the heat generated during re-entry is immense. Engineers have had to design advanced thermal protection systems and landing mechanisms to ensure that the booster can return to Earth safely.

Additionally, the Raptor engines that power both the *Super Heavy* booster and the *Starship* spacecraft are among the most complex rocket engines ever built. These engines use liquid methane and liquid oxygen as propellants, which allows for greater efficiency and the potential for refueling on other planets, such as Mars. However, the complexity of these engines has also led to development delays and technical issues.

Despite these challenges, SpaceX has made remarkable progress. The *Starship* prototypes have achieved successful high-altitude flights and landings, and the *Super Heavy* booster is undergoing extensive testing. Each test brings the company closer to realizing its goal of a fully reusable space transportation system.

The Role of Elon Musk's Ambition

Central to the development of *Starship* and *Super Heavy* is the unwavering ambition of Elon Musk. Musk's vision of making humanity a multi-planetary species has driven SpaceX's rapid advancements in rocket technology. His belief in the need for reusability to reduce the cost of space travel has been the guiding principle behind the development of *Super Heavy* and *Starship*.

Musk has repeatedly stated that his ultimate goal is to establish a self-sustaining colony on Mars. He views this as essential for the long-term survival of humanity, particularly in light of potential existential threats such as climate change, nuclear war, or asteroid impacts. For Musk, the ability to travel to and colonize other planets is not just a luxury or a scientific curiosity—it is a necessity.

This ambition has fueled SpaceX's willingness to take risks and innovate in ways that traditional aerospace companies have been unwilling or unable to do. Musk's leadership has allowed SpaceX to move quickly, often bypassing the bureaucratic hurdles and slow decision-making processes that characterize government agencies and larger corporations.

The Role of NASA and Commercial Partnerships

While Elon Musk's vision has been the driving force behind the development of *Starship* and *Super Heavy*, SpaceX has not achieved its success in isolation. The company has benefited from partnerships with government agencies, particularly NASA, as well as private companies and international partners.

NASA has played a critical role in SpaceX's development, both as a customer and as a partner in innovation. The space agency has awarded SpaceX multiple contracts, including those for launching cargo to the ISS and, more recently, for the Artemis missions, which aim to return humans to the Moon by 2026. NASA selected *Starship* as the lunar lander for these missions, recognizing its potential to revolutionize space travel.

These partnerships have provided SpaceX with the funding and resources necessary to accelerate the development of *Starship* and *Super Heavy*. In return, NASA has benefited from SpaceX's rapid innovation and cost-effective solutions. The partnership has created a new model for collaboration between government and private industry in space exploration.

The Economic Impact of Reusability

The development of *Starship* and *Super Heavy* has far-reaching economic implications. Traditional space missions are extraordinarily expensive, with launch costs often running into the hundreds of millions of dollars. The primary reason for these high costs is the need to build new rockets for each mission. With reusability, however, the cost of launching spacecraft can be reduced dramatically.

SpaceX has already demonstrated this with the *Falcon 9*, which can be launched and landed multiple times. The development of *Starship* and *Super Heavy* takes this a step further, with the potential to reduce the cost of space travel by an order of magnitude. This reduction in cost opens up new opportunities for commercial space ventures, from satellite deployment to space tourism.

Additionally, the reusability of *Starship* and *Super Heavy* has the potential to create new industries and jobs. The need for new infrastructure to support frequent space missions, such as refueling stations, habitats, and other space-based facilities, will drive economic growth and innovation.

The Path Forward

The development of *Starship* and *Super Heavy* is far from complete, but SpaceX is making rapid progress. The company continues to test and refine its designs, with the ultimate goal of making space travel routine and accessible. The successful return of the *Starship Engine Core* to the launch tower is just one of many milestones on this journey.

In the coming years, SpaceX will focus on scaling up its operations, with the goal of launching regular missions to the Moon and Mars. The success of these missions will depend on the continued development of reusable technology and the ability to refuel spacecraft in orbit. SpaceX is already working on solutions to these challenges, including the development of orbital refueling stations and new propulsion systems.

The future of space travel depends on the continued success of reusable spacecraft, and SpaceX is leading the charge. From the early concepts of Falcon rockets to the creation of the Starship and Super Heavy, Elon Musk and his team at SpaceX have redefined what is possible in space exploration.

The Starship and Super Heavy represent more than just technological advancements; they embody a vision for the future. A future where humans are no longer confined to

Earth but are free to explore and inhabit other planets. The successful return of the Starship engine core to its base is just the beginning of this bold journey. As SpaceX continues to innovate, the world watches with anticipation, knowing that humanity's future in space is within reach. The next chapters will explore this journey in more detail, covering the technology, challenges, and future possibilities that lie ahead.

CHAPTER 2

Reusability in Space Travel: Why It Matters

The concept of reusable rockets has become one of the most groundbreaking advancements in the modern era of space exploration, a technology that has redefined the economics, accessibility, and future possibilities of human ventures into space. Central to this transformation is SpaceX, the private space exploration company founded by Elon Musk in 2002, which has pushed the boundaries of what reusable technology can achieve. The development and implementation of reusable rockets, starting with the *Falcon 9* and culminating in the fully reusable *Starship*, represent a monumental leap forward in making space travel more sustainable, affordable, and practical.

This chapter will explore why reusability in space travel is so vital, examine how *Falcon 9* revolutionized the aerospace industry, and delve into how *Starship* is poised to take reusable technology to the next level, potentially unlocking the door to interplanetary colonization.

The Historical Context: Single-Use Rockets and Their Limitations

Historically, space travel has been an expensive and wasteful endeavor. Traditional rockets were built for single-use missions, meaning they were discarded after one flight, resulting in enormous costs. A typical space mission would see a rocket's various stages either falling into the ocean or burning up in the atmosphere, never to be used again. This method of launch was akin to throwing away a plane after every flight, a comparison often made by Elon Musk when discussing the inefficiency of traditional aerospace methods.

These single-use rockets made space travel prohibitively expensive for all but the wealthiest governments and a handful of private enterprises. For example, during the Apollo missions, NASA's Saturn V rocket, which carried astronauts to the Moon, was entirely expendable. Every launch consumed a massive amount of resources, including fuel, materials, and labor, which could never be recovered. The result was that only a few missions could be undertaken, limiting the scope and frequency of space exploration.

SpaceX's vision was to change this paradigm entirely, focusing on the principle of reusability to drastically reduce the cost of space travel. Reusable rockets, which can launch, land, and be flown again with minimal refurbishment, offer the promise of making space accessible to more nations, private companies, and even tourists. This approach also aligns with Musk's larger goal: making humanity a multi-planetary species by colonizing Mars, a venture that would be financially impossible without significant cost reductions.

Falcon 9: The First Step Towards Reusability

SpaceX's *Falcon 9* rocket became the company's first major success in the field of reusability. Introduced in 2010, the *Falcon 9* was designed with a reusable first stage, making it a breakthrough in aerospace technology. After launching payloads into orbit, the rocket's first stage would detach, flip, and return to Earth, landing vertically on a designated platform—either on a drone ship at sea or on solid ground near the launch site.

The development of this system was not without its challenges. The first attempts to land the *Falcon 9* stage were unsuccessful, with many boosters crashing or exploding upon impact. However, SpaceX persisted, refining its techniques with each test. In December 2015, SpaceX successfully landed the first stage of a *Falcon 9* rocket after launching it into orbit, a historic moment that demonstrated the feasibility of reusable rockets.

This breakthrough marked a turning point in space exploration. The ability to recover and reuse rockets significantly lowered the cost of each launch. To illustrate, before the advent of reusability, launching a satellite or spacecraft into orbit could cost tens or even hundreds of millions of dollars. The traditional expendable rocket models made these costs unavoidable. However, with the *Falcon 9*, the cost of space access plummeted. Reusing the first stage of the rocket meant that much of the expense associated with building new rocket stages for every launch was eliminated.

SpaceX's success with *Falcon 9* had several far-reaching consequences:

1. **Cost Reduction**: Reusability slashed launch costs. According to SpaceX, a fully expendable launch using the *Falcon 9* costs around $62 million. However, reusing the first stage could lower the cost by as much as 30%, making space launches more economically feasible.

2. **Increased Launch Frequency**: By reusing the rockets, SpaceX could launch more frequently. This increase in launch cadence is crucial for commercial satellite deployment, national security missions, and scientific research.

3. **A Competitive Market**: SpaceX's reusable rocket technology disrupted the traditional aerospace industry. Competitors such as the United Launch Alliance (ULA) and European space company Arianespace faced pressure to innovate and reduce costs. SpaceX became the go-to launch provider for a wide range of clients, including NASA, private satellite companies, and international governments.

4. **Environmental Impact**: Traditional rockets that are discarded after a single use leave behind debris, whether in the ocean or in Earth's orbit. Reusable rockets reduce the need for constant manufacturing and mitigate environmental waste.

These advantages underscored the critical importance of reusability in space travel, and SpaceX's *Falcon 9* became the first major step towards transforming space exploration from a rare and costly venture into a regular and sustainable industry.

The Starship System: Taking Reusability to the Next Level

While the *Falcon 9* represented a breakthrough in reusable technology, SpaceX had much grander ambitions. The *Starship* system, which consists of the *Starship* spacecraft and the *Super Heavy* booster, is designed to take the concept of reusability even further. Unlike the *Falcon 9*, which is only partially reusable (the second stage remains expendable), both stages of the *Starship* system are fully reusable.

The Super Heavy Booster

The *Super Heavy* booster is a massive rocket designed to launch the *Starship* spacecraft into orbit. Equipped with 33 Raptor engines, the *Super Heavy* is the most powerful rocket ever developed, capable of producing more than twice the thrust of NASA's Saturn V rocket, which carried humans to the Moon. After propelling the *Starship* spacecraft out of the atmosphere, the *Super Heavy* booster detaches and returns to Earth.

Like the *Falcon 9*, the *Super Heavy* booster is designed to land vertically after reentry, either on a floating platform at sea or directly at the launch site. However, SpaceX has introduced an even more ambitious system with the *Super Heavy* booster: rather than landing on legs, the booster is intended to be caught by the launch tower's mechanical arms, known as "chopsticks." This approach aims to reduce the wear and tear on the booster by eliminating the need for landing legs and allowing for faster turnaround times between launches.

The ability to rapidly reuse the *Super Heavy* booster is a key component of SpaceX's vision for routine space travel. The goal is to enable multiple launches and landings with minimal refurbishment, dramatically reducing the cost and time required for space missions.

The Starship Spacecraft

The *Starship* spacecraft, which sits atop the *Super Heavy* booster, is designed to be fully reusable as well. After the booster propels it into space, *Starship* can continue to its destination—whether that is low Earth orbit (LEO), the Moon, Mars, or beyond. Upon completing its mission, the spacecraft is capable of re-entering the Earth's atmosphere and landing vertically, much like the *Falcon 9*.

What sets the *Starship* spacecraft apart from previous designs is its versatility and scalability. The *Starship* is intended to serve a wide variety of missions, from launching satellites and space station crew rotations to deep space exploration and colonization efforts. The spacecraft is designed to carry up to 100 passengers, along with cargo, making it the most ambitious space vehicle ever conceived.

Moreover, the *Starship* spacecraft can be refueled in orbit, which is a game-changing capability for deep-space missions. By launching a fully fueled *Starship* into orbit and then sending a tanker version to replenish its fuel reserves, SpaceX can extend the range of the spacecraft far beyond Earth's orbit. This refueling process is essential for missions to the Moon, Mars, and beyond, as it allows the *Starship* to carry enough fuel to reach distant destinations without needing to refuel on the surface.

Why Reusability Matters for Interplanetary Travel

The development of fully reusable rockets like *Starship* and *Super Heavy* is critical to achieving SpaceX's long-term goal of making humanity a multi-planetary species. Sending humans to Mars—or any other planet—requires an unprecedented amount of resources, both in terms of technology and funding. Without reusability, the cost of launching and sustaining missions to other planets would be astronomical, far beyond the reach of any government or private company.

Reusability addresses this challenge by significantly reducing the cost of each launch. By reusing the same rocket for multiple missions, SpaceX can lower the per-launch cost, making it feasible to send regular missions to Mars. This cost reduction is not only important for exploration but also for the establishment of a self-sustaining colony on another planet. A Mars colony would require a constant supply of resources—water, food, equipment, and people—and reusable rockets would make this possible.

Furthermore, reusability also allows for more frequent missions, which is crucial for interplanetary colonization. To establish a permanent human presence on Mars, for example, we would need to send a continuous stream of supplies and personnel. Traditional rockets, which can only be used once, would make this process slow, expensive, and unsustainable. Reusable rockets, however, enable rapid and repeated launches, allowing for the steady flow of resources needed to support life on another planet.

Exp#### Why Reusability Matters for Interplanetary Travel (Continued)

The development of fully reusable rockets like *Starship* and *Super Heavy* is key to unlocking the future of interplanetary travel, especially for ambitious goals like Mars colonization. Without the capacity to reuse spacecraft and rocket boosters, each mission to a distant planet would require enormous financial investments and extensive resource consumption. With traditional, single-use rockets, the costs associated with launching, transporting, and sustaining life on another planet become prohibitive.

Reusability solves several critical challenges related to deep space exploration:

1. **Cost Efficiency**: By reusing the rocket components, the cost of each launch drops significantly. This affordability is vital for long-term space missions, especially when thinking of sending not just astronauts, but also infrastructure, food, equipment, and technology to support an off-world colony. The vision of establishing a permanent human settlement on Mars requires frequent deliveries of goods and people, which only reusable rockets can make feasible.

2. **Mission Frequency**: For interplanetary travel, it's not just about reaching a planet once; the goal is sustained travel. SpaceX's vision for Mars includes sending many missions over the course of several years to establish a functioning colony. Reusable technology allows for regular, scheduled flights, much like how airliners work. This increased frequency of missions enables faster development of an interplanetary supply chain.

3. **Scalability**: The ability to scale operations is crucial. With reusable rockets, SpaceX can scale up its operations to handle a growing number of missions to Mars, and eventually other planets. This scalability will allow the company to meet the demands of scientific research, space tourism, and government or commercial missions in the future.

4. **Environmental Benefits**: Traditional rocket launches involve discarding parts of the spacecraft, leaving debris either in space or in Earth's oceans. Reusable rockets reduce this waste, contributing to a more sustainable approach to space travel. As humanity looks toward the future, finding ways to explore space without generating excessive debris is becoming a growing concern. SpaceX's reusable technology addresses some of these environmental challenges.

With these key advantages, reusable rockets represent the future of space exploration, unlocking new opportunities for travel to the Moon, Mars, and potentially even farther destinations. As SpaceX refines the *Starship* system and begins using it for regular missions, the dream of sustained human presence on another planet will move closer to reality.

SpaceX's Legacy and the Road Ahead

SpaceX's commitment to reusability, showcased through the *Falcon 9* and now the *Starship*, has not only disrupted the aerospace industry but also set a new standard for space travel. The implications of reusable rockets extend far beyond cost savings; they redefine humanity's relationship with space and our potential to expand beyond Earth.

As we look toward the future, the continued success of reusable technology will open the door to new industries, such as space tourism, asteroid mining, and space-based manufacturing. Elon Musk's ultimate goal of establishing a colony on Mars is an ambitious one, but it is no longer science fiction. It is within the realm of possibility, thanks in large part to the technological advancements in reusability.

In the coming years, as SpaceX continues to refine and perfect the *Starship* system, we can expect to see more frequent missions to space, carrying not just astronauts, but also everyday citizens, scientists, and entrepreneurs. The future of space is one where reusability is the norm, and with each successful landing of a reusable rocket, we take one step closer to making space travel accessible to all.

The future of space exploration, colonization, and commercialization hinges on the success of reusable technology. And while there are still challenges to overcome, SpaceX has proven time and again that with persistence, innovation, and a relentless drive to push the boundaries of what's possible, humanity's future in space is brighter than ever.

CHAPTER 3
The Science of Controlled Landings

SpaceX has consistently pushed the boundaries of space exploration with its innovations in rocket technology, and one of the most impressive feats has been the ability to return rocket boosters to Earth in a controlled, precise manner. The *Super Heavy* booster, part of SpaceX's *Starship* launch system, is designed to return to its launch tower after propelling the *Starship* spacecraft into space. What sets this apart from previous rocket landings is the use of the "chopsticks" mechanism—a sophisticated and highly complex system that allows the booster to be caught mid-air by mechanical arms attached to a tower, instead of landing on the ground or sea.

This chapter delves into the technology and precision required for these controlled landings, the challenges SpaceX faced, and the engineering breakthroughs that made it possible. From cutting-edge navigation systems to powerful engines and automated landing procedures, this chapter will explain how SpaceX has turned rocket landings into a science.

The Basics of Rocket Landing: From Falcon 9 to Super Heavy

Before diving into the intricacies of *Super Heavy* landings, it is important to understand how SpaceX has been refining the technology of rocket recovery. It began with the *Falcon 9*, the workhorse of SpaceX, which was the first rocket to successfully return to Earth after launching payloads into space. The first stage of the *Falcon 9* booster was designed to return to the ground vertically, landing either on a ground pad or a drone ship in the ocean.

To achieve this, SpaceX developed a series of technological advancements, including:

1. **Grid Fins**: These aerodynamic surfaces deploy as the booster re-enters the atmosphere, allowing it to steer itself toward the landing zone. The grid fins provide precise control, allowing the rocket to make small adjustments in its trajectory during descent.

2. **Merlin Engines**: The *Falcon 9* uses Merlin engines that can be reignited during descent to control the speed of the rocket and to provide the thrust necessary to slow it down before landing.

3. **Autonomous Navigation**: The rocket is equipped with a highly advanced navigation system that uses GPS, radar, and sensors to adjust the landing trajectory in real-time, ensuring that it hits the landing pad with pinpoint accuracy.

These key innovations laid the foundation for the more complex landings that would follow with the *Super Heavy* booster. However, while the *Falcon 9* booster lands on the ground or a floating drone ship, the *Super Heavy* aims to be caught mid-air using mechanical arms—requiring even greater precision.

The "Chopsticks" Mechanism: A New Approach to Rocket Recovery

The *Super Heavy* booster, the largest and most powerful rocket SpaceX has ever developed, required an entirely new approach to landing. Due to its size and power, traditional landing legs (used in the *Falcon 9*) would be too bulky, heavy, and prone to damage upon repeated use. Instead, SpaceX introduced the "chopsticks" system—a mechanism that catches the booster in mid-air as it returns to the launch site.

The *chopsticks* system is part of a large structure called *Mechazilla*, a towering launch and recovery tower equipped with long mechanical arms. These arms extend out from the tower and are designed to catch the *Super Heavy* booster as it descends, guiding it safely back into position for future use. The decision to use such a system was driven by several factors:

1. **Weight Savings**: Removing landing legs from the booster reduces its overall weight, allowing more payload to be carried into space.

2. **Rapid Reusability**: By catching the booster with the arms, SpaceX can return it to the launch pad much faster than if it were to land traditionally. The booster can then be refueled and prepared for another launch in a much shorter time frame, which is essential for Musk's vision of frequent, low-cost space missions.

3. **Landing Precision**: The *chopsticks* system demands an even higher level of precision than traditional landings. The booster must descend at exactly the right angle and speed to align with the tower's arms, which are capable of making real-time adjustments to position themselves under the descending rocket.

The Science of Controlled Descent

The success of the *chopsticks* mechanism depends on the booster's ability to perform a controlled descent from space. After the booster separates from the *Starship* spacecraft, it begins its journey back to Earth. Several key technologies are involved in ensuring that this descent is precise and safe:

1. **Grid Fins for Aerodynamic Control**: Like the *Falcon 9*, the *Super Heavy* booster is equipped with grid fins that deploy after separation. These fins act as control surfaces, allowing the booster to steer itself during descent. Given the massive size and weight of the *Super Heavy*, these grid fins are much larger and must withstand greater forces than those on the *Falcon 9*.

2. **Raptor Engines for Thrust Control**: The *Super Heavy* booster is powered by Raptor engines, which use liquid methane and liquid oxygen as propellants. During descent, the engines are reignited to slow the booster's velocity, controlling the speed at which it approaches the landing zone. The Raptor engines are designed to throttle down smoothly, providing fine-tuned control over the booster's descent speed.

3. **Precision Guidance and Navigation**: One of the most important aspects of a controlled landing is the booster's ability to accurately determine its position and trajectory. The *Super Heavy* booster is equipped with a suite of navigation systems, including GPS, radar, and onboard sensors, that provide real-time data on the rocket's altitude, speed, and position. These systems feed data into the rocket's flight computer, which constantly adjusts the booster's trajectory to ensure it lines up perfectly with the *Mechazilla* tower.

4. **Autonomous Landing Algorithms**: The *Super Heavy* booster operates autonomously during descent, relying on advanced algorithms to make split-second decisions about how to adjust its course. These algorithms factor in wind speed, atmospheric conditions, and the rocket's current trajectory to calculate the precise maneuvers needed to align the booster with the *chopsticks* system.

Challenges in Developing the "Chopsticks" System

While the idea of catching a massive rocket booster with mechanical arms sounds simple in theory, the execution is anything but. SpaceX faced several major challenges during the development of the *chopsticks* system:

1. **Structural Integrity**: The *chopsticks* need to support the full weight of the *Super Heavy* booster, which weighs several hundred tons. The arms must be incredibly strong, yet flexible enough to adjust to the booster's movements during descent. Engineers had to carefully design the arms to ensure they could withstand the immense forces generated by the booster's weight and velocity.

2. **Real-Time Adjustments**: The *chopsticks* mechanism must be able to make real-time adjustments as the booster approaches. This requires the use of powerful sensors and actuators that can move the arms precisely and quickly. Any delay or miscalculation could result in a failed catch, leading to the potential destruction of the booster.

3. **Landing Accuracy**: The booster must land within an extremely tight margin of error to be successfully caught by the *chopsticks*. SpaceX engineers developed advanced landing algorithms that allow the booster to adjust its position and orientation with extreme precision. The grid fins and engines must

work in perfect harmony to ensure the booster lands exactly where it needs to.

4. **Heat and Stress Management**: During re-entry, the *Super Heavy* booster experiences immense heat and pressure as it passes through the Earth's atmosphere. This heat is enough to damage or destroy traditional materials, so SpaceX developed advanced thermal protection systems to shield the booster. These systems must protect the rocket's sensitive electronics and structure while still allowing it to maneuver and land safely.

Engineering Breakthroughs That Made It Possible

Several key engineering breakthroughs were necessary to make the *chopsticks* system and the controlled landings of the *Super Heavy* booster a reality. These breakthroughs not only addressed the challenges mentioned above but also pushed the boundaries of what is possible in aerospace engineering.

Raptor Engines: The Heart of Precision Landing

At the core of the *Super Heavy*'s success is the Raptor engine, an advanced rocket engine that uses liquid methane and liquid oxygen (also known as *methalox*) as propellants.

The choice of methane over traditional rocket fuels like RP-1 (a refined form of kerosene) offers several advantages for both space travel and reusability:

- **Higher Efficiency**: The Raptor engines are more efficient than SpaceX's Merlin engines, delivering greater thrust per unit of fuel. This efficiency is critical for the booster's ability to make fine adjustments during descent and for achieving the high power required to lift the massive *Super Heavy* and *Starship* system.

- **Throttle Control**: One of the unique features of the Raptor engines is their ability to throttle down to very low levels, allowing for precise control during landing. This throttle ability is crucial when landing a rocket as large and powerful as *Super Heavy*. The engines must provide just enough thrust to slow the rocket, but not so much that it prevents the booster from descending.

- **Refuelability on Mars**: A long-term goal of the *Starship* program is to enable missions to Mars, and the choice of methane as a fuel is strategic. Methane can potentially be produced on Mars using a process called in-situ resource utilization (ISRU), which involves extracting carbon dioxide from the Martian atmosphere and combining it with hydrogen to create methane. This capability is critical for long-term missions to other planets, where refueling must occur on the surface.

Grid Fins: Perfecting Rocket Steering

One of the most visible parts of SpaceX's landing technology is the use of grid fins, a technology that has evolved significantly from the days of the *Falcon 9*. These fins act as aerodynamic control surfaces that allow the booster to steer itself as it descends through the atmosphere. In earlier rockets, parachutes or thrusters were used to attempt recovery, but grid fins provide far more control and precision.

- **Material Strength**: SpaceX engineers had to develop new materials and manufacturing processes for the grid fins to withstand the extreme forces of re-entry. These fins are exposed to incredible heat and pressure as the booster re-enters the atmosphere, and they must maintain their shape and effectiveness throughout the descent. The latest generation of grid fins are made of titanium, a material that offers the strength and heat resistance necessary for repeated use.

- **Advanced Aerodynamics**: The shape and configuration of the grid fins have been refined over the years to improve their ability to steer the booster. These fins provide lift and drag, allowing the rocket to make sharp turns and adjust its trajectory with a high degree of precision. The fins operate in tandem with the rocket's engines, providing additional control during descent.

- **Real-Time Adjustment**: The grid fins are controlled by actuators that adjust their position in real-time based on data from the rocket's navigation system. These actuators are capable of making small, precise movements, ensuring that the rocket remains on course during the descent. This level of control is critical for aligning the rocket with the *chopsticks* system during the final stages of landing.

Automated Precision: The Role of AI and Machine Learning

One of the unsung heroes in the development of controlled landings is the software that governs the booster's descent. SpaceX has incorporated advanced artificial intelligence

(AI) and machine learning algorithms into the rocket's guidance system, allowing the booster to make real-time adjustments to its trajectory. These algorithms are constantly learning and improving, based on data from previous flights.

- **Autonomous Landing**: The booster operates autonomously during its descent, relying on its AI-based navigation system to make thousands of calculations per second. This system takes into account variables such as wind speed, atmospheric conditions, and the booster's current position and velocity. By continuously adjusting its course, the booster ensures that it remains aligned with the *chopsticks* for a successful catch.

- **Flight Data Analysis**: Each time a booster lands, SpaceX collects a massive amount of data. This data is fed into machine learning models that are used to refine the rocket's landing algorithms. Over time, these models improve the booster's ability to handle different landing conditions, increasing the chances of a successful recovery on future missions.

- **Sensor Fusion**: The booster's navigation system uses a technique called sensor fusion, which combines data from multiple sources, including

GPS, radar, and onboard sensors. By integrating these data streams, the rocket can build a detailed, real-time picture of its surroundings, allowing for more accurate landings.

The Challenges of Heat and Vibration

Rocket landings are not just about precision; they are also about surviving the extreme conditions of re-entry. When a rocket descends back into Earth's atmosphere, it encounters tremendous heat and vibration, which can damage sensitive electronics and mechanical components.

- **Thermal Protection**: SpaceX engineers had to develop advanced thermal protection systems to shield the booster from the intense heat of re-entry. The *Super Heavy* booster is equipped with a heat shield that dissipates heat and protects the internal components of the rocket. This heat shield must be lightweight, durable, and capable of withstanding multiple re-entries.

- **Vibration Dampening**: The booster experiences intense vibrations during both launch and landing. These vibrations can cause damage to sensitive components, such as the engines and navigation system. To mitigate this, SpaceX developed vibration dampening systems that absorb and reduce the impact of these forces, ensuring that the booster remains functional during descent.

The Future of Controlled Landings

While the *Super Heavy* booster and *chopsticks* system represent the cutting edge of rocket recovery, SpaceX is constantly looking for ways to improve. One area of focus is the refinement of the landing process to make it even more efficient and reliable. SpaceX aims to reduce the turnaround time between launches, allowing the same booster to be used multiple times in rapid succession.

- **Improved Landing Algorithms**: SpaceX is working on refining the landing algorithms to further increase precision. By improving the accuracy of the rocket's descent, SpaceX can reduce wear and tear on the booster, extending its operational lifespan.

- **Faster Turnaround**: One of SpaceX's long-term goals is to achieve rapid reusability, where a booster can be launched, recovered, and relaunched within 24 hours. The *chopsticks* system is a key part of this strategy, as it allows the booster to be caught and returned to the launch pad quickly, eliminating the need for lengthy refurbishment processes.

- **Expanding Reusability**: In the future, SpaceX hopes to extend the principles of reusability beyond just the booster. The *Starship* spacecraft itself is designed to be fully reusable, capable of landing on the Moon or Mars and returning to Earth. By developing reusable spacecraft, SpaceX aims to make space travel more sustainable and affordable, paving the way for long-term missions to other planets.

The science of controlled landings has evolved from a futuristic dream into a practical reality, thanks to SpaceX's relentless pursuit of innovation. The development of the *Super Heavy* booster and the *chopsticks* system represents a major leap forward in rocket recovery technology. By perfecting the art of catching rockets mid-air, SpaceX has set the stage for a new era of rapid, reusable space travel.

As SpaceX continues to refine its landing systems, the possibilities for space exploration expand. The ability to recover and reuse rockets will make space more accessible, reducing costs and increasing the frequency of missions. The *Super Heavy* booster is just the beginning—future innovations in controlled landings will open the door to even greater achievements, from regular lunar landings to the eventual colonization of Mars.

In the coming years, SpaceX's controlled landing technology will likely become the standard for all space missions, both governmental and private. The precision and engineering required to land a rocket in such a controlled manner have set a new benchmark for the aerospace industry. As we look to the future, the science of controlled landings will play a central role in humanity's efforts to explore, settle, and thrive beyond Earth.

CHAPTER 4

Testing and Trials: The Journey to Success

SpaceX's journey to the successful catch of the *Super Heavy* booster using the *Mechazilla* tower and its "chopsticks" mechanism has been an intense process of iteration, experimentation, failure, and incremental improvement. This chapter will explore the extensive series of tests and trials that led to the technological marvel that is today's *Starship* and *Super Heavy* system. SpaceX has adopted a philosophy of rapid prototyping, with its famous motto of "fail fast, learn faster." This iterative approach has been instrumental in transforming experimental concepts into operational breakthroughs.

From early explosive failures to triumphant high-altitude tests, the road to success was paved with setbacks, but each failure provided a wealth of information that allowed the engineers at SpaceX to refine the *Starship* and *Super Heavy* systems. This chapter chronicles the crucial milestones, lessons learned, and the technological breakthroughs that came from each test and trial.

Early Testing: The Foundations of Reusability

Before SpaceX began working on the *Starship* and *Super Heavy*, the foundation of reusability was laid with the development of the *Falcon 9* rocket. The company had already demonstrated that the first stage of a rocket could be recovered and reused—a monumental achievement in itself. These early landings of the *Falcon 9* booster provided critical insights into the complexities of rocket recovery, which would later be applied to the much larger and more powerful *Super Heavy*.

The *Falcon 9* landings were far from perfect at first. Early attempts to recover the booster led to several failed landings, with boosters either crashing into the ocean or landing too hard and being destroyed. However, each failure gave SpaceX valuable data, and after each failed test, the company made adjustments. The inclusion of grid fins, more precise navigation systems, and improved landing legs all played crucial roles in the ultimate success of *Falcon 9*. These same innovations would become the cornerstones of the *Super Heavy* booster's recovery system.

The Early Prototypes: Star hopper and Starship Mk1

After proving the feasibility of booster recovery with the *Falcon 9*, SpaceX began working on the next phase of its reusability goals—developing a fully reusable system capable of taking humans to the Moon, Mars, and beyond. The first iteration of this new system came in the form of *Star hopper*, an early prototype for *Starship*. *Star hopper* was a scaled-down, single-engine version of what would eventually become the *Starship* vehicle.

Star hopper was designed to test the new Raptor engine, which is far more powerful than the Merlin engines used on the *Falcon 9*. The Raptor engine uses methane and liquid oxygen as fuel, which allows for greater efficiency and the possibility of refueling on Mars, where methane can be synthesized from the Martian atmosphere. *Star hopper* successfully completed several low-altitude "hops" in 2019, reaching heights of up to 150 meters before landing safely back on the ground.

While *Star hopper* never flew to space, its successful tests provided valuable data on the performance of the Raptor engine, the control systems, and the landing mechanisms that would be used on larger vehicles. These early flights also demonstrated that the core concept of vertical landing could be scaled up for larger, more powerful rockets.

Next came the *Starship Mk1* prototype, the first full-scale test article for SpaceX's future interplanetary spacecraft.

Starship Mk1 was intended to reach higher altitudes and test more advanced systems, but its journey was cut short when it exploded during a pressurization test in November 2019. While the explosion was a setback, it provided important insights into the structural integrity of the vehicle and led to design improvements that were implemented in subsequent versions of *Starship*.

SN-Series: High-Altitude Flights and Explosive Landings

Following the failure of *Starship Mk1*, SpaceX moved forward with the SN (serial number) series of prototypes. Each prototype in the SN-series was designed to test different aspects of the *Starship* system, from the Raptor engines to the aerodynamic control surfaces and the landing system.

- **SN5 and SN6**: These prototypes were stripped-down versions of the final *Starship* design, featuring only a single Raptor engine and no aerodynamic flaps. Both vehicles successfully completed low-altitude hops similar to *Star hopper*, reaching heights of 150 meters before landing. These tests helped validate the basic systems required for controlled flight and landing, but the real challenges were yet to come.

54

- **SN8**: The SN8 prototype was the first *Starship* to feature the full suite of Raptor engines, aerodynamic flaps, and a nosecone, making it the most complete test vehicle to date. In December 2020, SN8 performed a high-altitude flight, reaching an altitude of 12.5 kilometers (approximately 7.8 miles). The vehicle performed a dramatic belly-flop maneuver during its descent, using its flaps to control its orientation and speed. However, during the final moments of landing, SN8 experienced a failure in one of its Raptor engines, causing it to crash and explode upon impact.

Despite the fiery end to the test, SN8 was a major success. It demonstrated that the *Starship* could perform complex maneuvers and maintain stability during descent, which is critical for landing on other celestial bodies like the Moon and Mars. The failure during landing provided SpaceX with data on the performance of the Raptor engines during the final moments of descent, leading to improvements in future tests.

- **SN9 and SN10**: Both SN9 and SN10 followed similar flight paths to SN8, with each vehicle reaching high altitudes and performing the belly-flop maneuver during descent. SN9 met a similar fate as SN8, crashing during the final moments of landing due to an engine failure. SN10, however, marked a significant milestone. The vehicle completed its high-altitude flight and belly-flop maneuver, and unlike its predecessors, it managed to land upright on the launch pad. Unfortunately, a small methane leak caused SN10 to explode a few minutes after landing, but the successful landing was a major breakthrough.

- **SN11 and Beyond**: SN11 flew in March 2021 but experienced an issue during re-entry, leading to an in-flight explosion before it could land. Despite this failure, each test provided invaluable data that SpaceX used to refine the *Starship* design.

The Evolution of the Super Heavy Booster: New Challenges

While the SN-series tests focused primarily on the *Starship* spacecraft itself, the *Super Heavy* booster was being developed in parallel. The *Super Heavy* booster, designed to provide the immense thrust needed to propel *Starship* into space, presented new challenges due to its size, weight, and complexity. With 33 Raptor engines, *Super Heavy* is the most powerful rocket booster ever built.

Testing the *Super Heavy* booster involved developing systems to control its reentry and landing. Like the *Falcon 9* booster, *Super Heavy* is designed to return to Earth after launching *Starship* into space. However, due to its massive size, landing legs were deemed impractical. Instead, SpaceX opted to develop the *Mechazilla* tower and its *chopsticks* mechanism to catch the booster mid-air.

The Development of the "Chopsticks" Catching Mechanism

The decision to use the *chopsticks* mechanism for catching the *Super Heavy* booster was a significant departure from traditional rocket recovery methods. While the *Falcon 9* booster lands vertically using landing legs, *Super Heavy* is designed to be caught mid-air by the mechanical arms of the *Mechazilla* tower. This innovative approach offers several advantages, including reducing the weight of the booster by eliminating landing legs and allowing for faster turnaround times between launches.

The development of the *chopsticks* system required extensive testing and engineering. The arms needed to be strong enough to support the weight of the booster, which weighs hundreds of tons, while also being flexible enough to make real-time adjustments during the catch. The booster itself had to be equipped with precise navigation systems to align with the tower's arms during descent.

SpaceX conducted several tests to perfect the *chopsticks* system, starting with small-scale prototypes and gradually working up to full-scale tests with the actual booster. These tests helped SpaceX refine the timing, positioning, and control systems needed to catch the booster successfully.

Lessons Learned: The Iterative Process

One of the key factors in SpaceX's success has been its willingness to embrace failure as a learning opportunity. Each test flight, regardless of whether it ended in success or failure, provided valuable data that was used to improve future designs.

For example, the failures of SN8 and SN9 highlighted issues with the Raptor engines during the landing burn. In response, SpaceX made several modifications to the engine's design, including improving the methane fuel system to prevent pressure drops during landing. Similarly, the explosion of SN10 after landing revealed vulnerabilities in the landing legs and fuel tanks, leading to further refinements.

The iterative nature of SpaceX's testing program allowed the company to rapidly develop and improve the *Starship* and

Super Heavy systems. By testing often and learning from each failure, SpaceX was able to make significant progress in a relatively short period of time.

The Breakthrough: Successful Catch of the Booster

After years of testing and incremental improvements, SpaceX achieved a major breakthrough with the successful catch of the *Super Heavy* booster using the *Mechazilla* tower's *chopsticks*. This marked a critical milestone in SpaceX's journey to develop fully reusable rockets, as it demonstrated that the successful catch proved that SpaceX could not only launch but also recover and reuse the largest and most powerful rocket booster ever built. The event was the culmination of years of testing, countless iterations, and innovative engineering.

The first successful catch of the *Super Heavy* booster happened during one of the later *Starship* test flights. The booster performed flawlessly during launch, propelling the *Starship* into orbit before detaching and beginning its descent back to Earth. As the booster approached the *Mechazilla* tower, the *chopsticks* system extended, positioning itself under the descending rocket. The booster's grid fins and Raptor engines worked together to guide the massive vehicle into position, and the *chopsticks* successfully caught the booster mid-air, securing it safely back at the launch pad.

This achievement represented a critical milestone not just for SpaceX, but for the future of reusable space travel. By perfecting the *chopsticks* mechanism, SpaceX demonstrated that it could recover and reuse the *Super Heavy* booster without the need for landing legs or extensive refurbishment. This breakthrough brings SpaceX one step closer to achieving its goal of rapid reusability, where boosters can be launched, caught, refueled, and relaunched within a matter of hours.

The Significance of the Successful Catch

The successful catch of the *Super Heavy* booster using the *chopsticks* system was a monumental achievement for several reasons:

1. **Cost Savings**: The ability to catch the booster mid-air eliminates the need for landing legs, which reduces the overall weight of the vehicle and increases payload capacity. It also reduces the cost of recovery, as the booster can be returned directly to the launch pad without the need for a lengthy refurbishment process.

2. **Faster Turnaround Times**: One of the key goals of SpaceX is to achieve rapid reusability, where rockets can be launched, recovered, and relaunched within a short period of time. The *chopsticks* system is a crucial part of this vision, as it allows the booster to be returned to the launch pad quickly and efficiently.

3. **Increased Launch Frequency**: By reducing the time and cost associated with rocket recovery, SpaceX can increase the frequency of its launches. This is critical for future missions to the Moon, Mars, and beyond, where regular supply runs and crewed missions will be necessary.

4. **Paving the Way for Interplanetary Travel**: The success of the *Super Heavy* booster and *Starship* system is a major step toward SpaceX's long-term goal of making humanity a multi-planetary species. By developing fully reusable rockets, SpaceX is dramatically reducing the cost and complexity of space travel, making it possible to envision regular missions to the Moon, Mars, and other destinations in the solar system.

The Road Ahead: Future Testing and Improvements

While the successful catch of the *Super Heavy* booster marked a significant milestone, SpaceX's journey is far from over. The company continues to test and refine the *Starship* and *Super Heavy* systems, with the ultimate goal of making space travel as routine as air travel.

Future tests will focus on improving the reliability of the *chopsticks* system and further reducing the turnaround time between launches. SpaceX is also working on the development of orbital refueling systems, which will allow *Starship* to refuel in space and extend its range for deep-space missions. These innovations are critical for the eventual goal of sending humans to Mars.

In addition, SpaceX continues to test the *Starship* spacecraft itself, with plans for crewed missions to the Moon as part of NASA's Artemis program and eventually missions to Mars. Each test provides valuable data that helps SpaceX refine its designs and move closer to its goal of making interplanetary travel a reality.

A New Era in Space Exploration

The journey to successfully catching the *Super Heavy* booster has been one of the most challenging and rewarding chapters in SpaceX's history. From the early days of *Star hopper* and low-altitude hops to high-altitude tests and explosive landings, each trial has brought SpaceX closer to achieving its vision of reusable, affordable, and sustainable space travel.

The successful catch of the *Super Heavy* booster using the *chopsticks* system represents a major milestone in the development of fully reusable rockets. By perfecting this technology, SpaceX has not only revolutionized the way we think about space travel but has also opened the door to new possibilities for human exploration of the Moon, Mars, and beyond.

As SpaceX continues to push the boundaries of what is possible, the future of space exploration looks brighter than ever. The lessons learned from the countless tests and trials leading up to this moment will pave the way for a new era in space travel, where rockets can be launched, recovered, and reused with unprecedented speed and efficiency. And with each new breakthrough, humanity moves one step closer to becoming a multi-planetary species.

CHAPTER 5

Chopsticks: The Game-Changing Landing Mechanism

The development of SpaceX's "chopsticks" landing mechanism is one of the most innovative and daring technological advancements in modern rocketry. This mechanism, part of the larger *Mechazilla* tower, allows the massive *Super Heavy* booster to be caught mid-air, offering a revolutionary alternative to traditional rocket landing techniques. It has the potential to transform the way reusable rockets are recovered and prepared for future missions, paving the way for rapid rocket reuse and lowering the costs of space travel.

In this chapter, we will take an in-depth look at the chopsticks mechanism, explore how it compares to conventional landing techniques, and analyze how it fits into the larger context of SpaceX's goal of enabling affordable, frequent space travel.

Traditional Landing Techniques: The Precursor to Chopsticks

To fully appreciate the significance of the chopsticks system, it's important to first understand the evolution of rocket landing technology. Traditional rockets, such as those used during the Apollo missions or by agencies like Russia's Roscosmos, were entirely expendable. After launch, rocket boosters would fall back to Earth, either burning up in the atmosphere or crashing into the ocean, making them impossible to recover and reuse. This process was extremely wasteful and expensive.

SpaceX began to revolutionize the industry by introducing the concept of reusability with its *Falcon 9* rocket, which became the first rocket capable of landing itself and being reused for future missions. The *Falcon 9*'s first stage booster is equipped with retractable landing legs and grid fins, which enable it to perform a controlled descent back to Earth after launch. The booster uses its engines to slow its descent and performs a final landing burn to touch down vertically either on solid ground or on a floating drone ship in the ocean.

This vertical landing system was a monumental leap forward in rocket technology, enabling SpaceX to dramatically lower the cost of space missions by reusing its boosters. However, this system still had its limitations. The landing legs added weight to the rocket, reducing payload capacity, and the process of recovering the booster from a remote ocean location was time-consuming and logistically challenging.

The Vision for Chopsticks and Rapid Reusability

As SpaceX looked ahead to the development of its next-generation *Super Heavy* booster and *Starship* spacecraft, CEO Elon Musk envisioned a future where rockets could be launched, recovered, and relaunched in a matter of hours—similar to the way airplanes are turned around between flights. This vision of rapid reusability required a new approach to rocket recovery, one that would eliminate the need for landing legs and ocean recoveries.

The chopsticks system was born out of this vision. Rather than landing the rocket on legs, the *Super Heavy* booster is caught mid-air by the mechanical arms of the *Mechazilla* tower. These arms, nicknamed "chopsticks" for their long, slender appearance, are designed to extend from the launch tower and grab the descending booster at specific attachment points. Once caught, the booster can be immediately repositioned and prepared for the next launch.

This system offers several key advantages over traditional landing techniques:

1. **Weight Savings**: By eliminating landing legs, the *Super Heavy* booster is lighter, which allows it to carry more payload into space. Every kilogram saved on the rocket's structure translates into more capacity for satellites, crew, cargo, or other mission-critical components.

2. **Speed of Recovery**: The chopsticks mechanism is designed to catch the booster at the launch site, eliminating the need for recovery operations at remote locations or on ocean drone ships. This enables the booster to be quickly refueled, checked, and prepared for another launch, greatly reducing the turnaround time between missions.

3. **Simplified Logistics**: Traditional recovery methods, especially those that involve landing at sea, require complex logistics, including the use of ships and helicopters to retrieve the booster and return it to the launch site. The chopsticks system simplifies this process by keeping the entire operation at the launch site.

4. **Reduced Wear and Tear**: Landing legs, while effective, are subject to wear and tear after multiple landings. By catching the booster mid-air, SpaceX

avoids the stresses that come from hard landings, reducing the need for maintenance and increasing the lifespan of each booster.

How the Chopsticks Mechanism Works

The chopsticks mechanism is integrated into the *Mechazilla* tower, a towering structure located at SpaceX's Star base in Boca Chica, Texas. The *Mechazilla* tower serves two primary purposes: it acts as both the launch tower for the *Super Heavy* booster and *Starship* spacecraft, and it is equipped with the chopsticks system to recover the booster after each launch.

Here's a breakdown of how the chopsticks mechanism works:

1. **Booster Descent**: After the *Super Heavy* booster completes its role in launching the *Starship* spacecraft into space, it detaches and begins its descent back to Earth. During descent, the booster uses grid fins to steer itself toward the launch site, making real-time adjustments to ensure it aligns with the *Mechazilla* tower.

2. **Chopsticks Deployment**: As the booster approaches the tower, the chopsticks extend from the tower and

position themselves under the descending booster. The chopsticks are designed to catch the booster at specific attachment points located near its upper section, where the structure is strongest.

3. **Real-Time Adjustments**: The chopsticks are equipped with sensors and actuators that allow them to make real-time adjustments during the catch. This is critical, as the booster is descending at high speed and must be caught with precision to avoid damage.

4. **Catching the Booster**: Once the booster is in position, the chopsticks close around the attachment points and catch the booster mid-air. The arms then guide the booster back into its launch position on the pad, where it can be secured for refueling and inspection.

5. **Rapid Turnaround**: Once the booster is caught and secured, the recovery process is complete. The booster can be prepared for its next launch with minimal downtime, enabling rapid reusability.

Engineering Challenges and Solutions

Developing the chopsticks system required overcoming several engineering challenges, as it represents a dramatic departure from traditional rocket recovery techniques. SpaceX's engineers had to design a system capable of withstanding the immense weight and forces generated by the *Super Heavy* booster, while also ensuring that the booster could be caught with pinpoint precision.

1. **Structural Strength**: The chopsticks are designed to catch a rocket booster that weighs several hundred tons, making structural strength a key concern. The arms must be strong enough to support the booster's weight and withstand the dynamic forces of its descent. To achieve this, the chopsticks are constructed from high-strength materials and are reinforced to ensure they can handle repeated use.

2. **Precision Alignment**: One of the most difficult aspects of the chopsticks system is the need for precise alignment between the booster and the arms. The booster must descend along a specific trajectory, and any deviation could result in a missed catch. SpaceX addressed this challenge by equipping the *Super Heavy* booster with advanced navigation systems, including GPS, radar, and grid fins, which allow the booster to steer itself with incredible accuracy during descent.

3. **Real-Time Adjustments**: To further ensure a successful catch, the chopsticks are equipped with sensors and actuators that allow them to make real-time adjustments. These adjustments are controlled by algorithms that take into account the booster's speed, position, and orientation. The system continuously analyzes data from the booster's sensors and makes minute adjustments to ensure that the arms are in the correct position to catch the booster.

4. **Safety and Redundancy**: Given the complexity and risks involved in catching a rocket mid-air, SpaceX has built multiple layers of safety and redundancy into the chopsticks system. This includes fail-safe mechanisms that can abort the catch if something goes wrong, as well as backup systems that ensure the arms can still function in the event of a technical failure.

Comparison to Traditional Landing Techniques

The chopsticks mechanism is a revolutionary departure from the traditional landing techniques used by SpaceX with the *Falcon 9* and other rockets in the aerospace industry. Here's how the chopsticks system compares to more conventional methods:

1. **Landing Legs vs. Mid-Air Catch**: The most obvious difference between the two systems is that the *Falcon 9* uses landing legs to touch down vertically on solid ground or a drone ship, while the *Super Heavy* booster is caught mid-air by the chopsticks. Landing legs add weight to the rocket, reducing its payload capacity, while the chopsticks system eliminates this weight, allowing for greater payload efficiency.

2. **Landing Sites**: Traditional landings often require large landing pads or ocean-based recovery platforms, which can complicate logistics and delay recovery times. The chopsticks system allows the booster to be caught directly at the launch site, reducing the need for complex recovery operations and speeding up the turnaround process.

3. **Turnaround Time**: One of the primary advantages of the chopsticks system is the speed with which the booster can be recovered and prepared for relaunch. Traditional landings often require the booster to be transported back to the launch site, inspected, and

refurbished before it can be reused. With the chopsticks system, the booster is caught, repositioned, and ready for inspection and refueling in a fraction of the time.

4. **Reusability**: Both systems are designed with reusability in mind, but the chopsticks mechanism takes this concept to the next level by minimizing wear and tear on the rocket. By catching the booster mid-air, SpaceX reduces the mechanical stresses associated with traditional landings, increasing the lifespan of each booster and allowing it to be reused more frequently.

The Future of Rapid Rocket Reuse The chopsticks mechanism is not just a technical achievement but also a foundational element of SpaceX's vision for rapid rocket reuse. SpaceX's long-term goal is to make space travel as routine and accessible as possible, with rockets that can be launched, recovered, and relaunched in a matter of hours. The chopsticks system is central to this vision, as it enables the *Super Heavy* booster to be caught, repositioned, and prepared for relaunch in record time.

1. **Quicker Turnaround for Frequent Launches**: The chopsticks system eliminates the need for time-

consuming recovery operations at sea or remote locations. By catching the booster at the launch site, SpaceX can immediately begin the process of preparing the rocket for its next mission. This capability is critical for future missions to the Moon, Mars, and beyond, where frequent launches will be necessary to transport cargo, crew, and supplies.

2. **Cost Reductions**: Rapid reusability is not just about speed—it's also about cost savings. The faster SpaceX can recover and relaunch its rockets, the lower the cost per mission becomes. By reducing the downtime between launches, SpaceX can conduct more missions with fewer rockets, lowering the overall cost of space travel.

3. **Paving the Way for Interplanetary Colonization**: SpaceX's ultimate goal is to enable humanity to become a multi-planetary species, with colonies on Mars and other planets. This vision will require a fleet of reusable rockets capable of frequent, cost-effective missions to transport people, cargo, and supplies across the solar system. The chopsticks system is a crucial part of this vision, as it enables

SpaceX to achieve the rapid reusability needed to make interplanetary colonization a reality.

A Game-Changing Innovation in Rocket Recovery

The development of the chopsticks mechanism represents a monumental leap forward in the science of rocket recovery. By catching the *Super Heavy* booster mid-air, SpaceX has not only solved the problem of heavy, cumbersome landing legs but also paved the way for faster, more efficient rocket reuse. This system is a key element of SpaceX's broader mission to make space travel affordable, sustainable, and frequent.

The chopsticks system is not just about recovering rockets—it's about enabling a future where space exploration is routine. Whether launching satellites, transporting astronauts to the International Space Station, or carrying humans to the Moon and Mars, the ability to rapidly recover and relaunch rockets is crucial for the next era of space exploration.

As SpaceX continues to refine the chopsticks mechanism and other aspects of its rocket technology, the possibilities for human expansion into space grow ever closer. This innovative approach to rocket recovery represents a game-changing moment in the history of space travel, and it will undoubtedly play a central role in humanity's journey to the stars.

CHAPTER 6

From Earth to Mars: Starship's Interplanetary Mission

SpaceX's *Starship* system, comprising the *Starship* spacecraft and *Super Heavy* booster, is at the heart of the company's vision for humanity's future in space. Elon Musk, CEO and founder of SpaceX, has long advocated for making life multi-planetary, with Mars being the primary target for human colonization. SpaceX's broader goals for *Starship* are ambitious: the development of a reusable spacecraft capable of transporting humans and cargo not just to Earth's orbit or the Moon, but also to Mars and beyond.

This chapter will explore SpaceX's interplanetary mission, focusing on the technical challenges of long-duration spaceflight and how reusable boosters like *Super Heavy* are essential to making this ambitious goal a reality. As humanity contemplates a future on other planets, *Starship* is positioned as the vehicle that can make this dream achievable.

SpaceX's Broader Goals: Moon and Mars Missions

SpaceX's aspirations for *Starship* encompass missions to the Moon and Mars, with the ultimate goal of enabling humanity to establish permanent settlements on these celestial bodies. SpaceX is already part of NASA's Artemis program, which aims to return humans to the Moon by 2026. The *Starship* spacecraft has been selected as the lunar lander for these missions, marking a critical step toward demonstrating the spacecraft's versatility and capability for interplanetary travel.

However, Mars remains the ultimate goal. The planet, though barren and inhospitable, has long been seen as the best candidate for a self-sustaining human colony. Mars' relatively close proximity to Earth, its day length (about 24.6 hours), and the availability of key resources, such as water ice and carbon dioxide, make it a compelling target for colonization. SpaceX's broader goals involve establishing a city on Mars that could eventually become self-sufficient, ensuring humanity's survival in the event of catastrophic events on Earth.

The Role of Starship in Moon and Mars Missions

The *Starship* spacecraft is designed to carry large payloads, up to 100 metric tons, to a variety of destinations, including

Earth's orbit, the Moon, and Mars. Its fully reusable design is critical to reducing the cost of space travel, making it possible to undertake multiple missions without the expense of building new spacecraft for each trip.

1. **Moon Missions**: SpaceX's involvement in NASA's Artemis program highlights the capability of *Starship* to carry astronauts and cargo to the Moon. As the lunar lander, *Starship* will deliver crew members from lunar orbit to the Moon's surface and back again. Its large payload capacity and ability to land on unprepared surfaces make it well-suited for the Moon, where it can transport habitats, scientific equipment, and other resources needed for long-term lunar exploration.

2. **Mars Missions**: Mars presents a far more challenging target than the Moon, not only because of its distance from Earth but also due to its thin atmosphere, harsh climate, and lack of readily accessible resources. SpaceX has designed *Starship* to address these challenges, focusing on reusability, payload capacity, and the ability to refuel on the surface of Mars using locally sourced propellant. The goal is to make round-trip missions to Mars feasible

by refueling *Starship* on the planet and sending it back to Earth, significantly lowering the cost and complexity of such missions.

Technical Challenges of Long-Duration Spaceflight

While *Starship* represents a significant technological advancement, there are still numerous challenges associated with long-duration spaceflight to Mars. These challenges include propulsion, life support, radiation exposure, and the psychological effects of prolonged isolation.

1. **Propulsion and Fuel**: One of the biggest challenges for interplanetary travel is propulsion. SpaceX has developed the Raptor engine, which is capable of using liquid methane and liquid oxygen as propellants. Methane, in particular, is crucial because it can potentially be produced on Mars using a process called in-situ resource utilization (ISRU). By extracting carbon dioxide from Mars' atmosphere and combining it with hydrogen, methane can be synthesized and used as fuel for the return trip to Earth. This reduces the need to transport large amounts of fuel from Earth, lowering mission costs and complexity.

2. **Life Support Systems**: Sustaining human life on long-duration missions to Mars will require advanced life support systems capable of recycling air, water, and food. SpaceX has not revealed all the details of its life support technology, but it is likely to incorporate advanced systems to filter and recycle air, generate oxygen, and purify water. These systems will need to function autonomously for extended periods, as resupply missions will be infrequent and costly.

3. **Radiation Protection**: Space radiation is one of the most significant threats to human health during long-duration space missions. Outside of Earth's protective magnetic field, astronauts are exposed to high levels of cosmic rays and solar radiation, which can increase the risk of cancer and other health problems. Developing effective shielding to protect astronauts during the trip to Mars, which can take six to nine months, is critical. Materials such as water, hydrogen-rich compounds, or advanced polymers may be used in *Starship*'s design to provide radiation protection.

4. **Psychological and Social Challenges**: Space travel, especially long-duration missions, poses psychological and social challenges that are often overlooked. The isolation, confinement, and distance from Earth can have profound effects on the mental health of astronauts. SpaceX will need to consider the psychological well-being of the crew during multi-month trips to Mars. Providing a comfortable living environment, opportunities for recreation, and effective communication with Earth will be essential to maintaining the crew's mental health and ensuring mission success.

5. **Landing and Surface Operations on Mars**: Landing on Mars presents unique challenges due to its thin atmosphere, which provides less friction for decelerating incoming spacecraft. SpaceX has designed *Starship* to perform a powered descent using its Raptor engines, similar to how the *Falcon 9* lands on Earth. However, Mars' atmosphere and gravity are different from Earth's, so SpaceX will need to conduct extensive testing to ensure safe landings. Once on Mars, *Starship* will also serve as a habitat for the crew until more permanent structures can be built.

Super Heavy: The Essential Role of Reusable Boosters

The *Super Heavy* booster plays an essential role in making interplanetary missions possible by providing the thrust needed to propel *Starship* out of Earth's gravity well and into space. With 33 Raptor engines, *Super Heavy* is the most powerful rocket booster ever built, capable of lifting the fully loaded *Starship* spacecraft into orbit. However, what makes *Super Heavy* truly revolutionary is its reusability.

1. **Lowering Launch Costs**: Reusability is the key to reducing the cost of space travel. Traditional rockets are single-use, with each mission requiring a new rocket to be built at great expense. By contrast, *Super Heavy* is designed to be fully reusable, capable of returning to Earth after each launch and being quickly prepared for the next mission. This drastically reduces the cost of launching large payloads into space, making frequent trips to the Moon and Mars economically viable.

2. **Chopsticks Recovery System**: One of the most innovative features of *Super Heavy* is its recovery system. Instead of landing on legs like the *Falcon 9*, *Super Heavy* will be caught mid-air by the mechanical arms of the *Mechazilla* tower, nicknamed

"chopsticks." This system eliminates the need for landing legs, reducing the overall weight of the booster and allowing for more efficient recovery. The rapid turnaround time provided by this system is essential for SpaceX's vision of launching multiple missions in quick succession.

3. **Launch Cadence and Payload Capacity**: The sheer size and power of *Super Heavy* allow it to carry the massive payloads needed for interplanetary missions. Whether transporting crew, cargo, or habitats, the ability to lift such large payloads is critical for establishing a sustained presence on Mars. The reusability of *Super Heavy* enables SpaceX to launch frequent missions, ensuring a steady supply of resources for future Mars colonies.

4. **Orbital Refueling**: One of the challenges of launching interplanetary missions from Earth is the amount of fuel required to escape Earth's gravity and reach distant planets like Mars. SpaceX's solution to this problem is orbital refueling. After *Super Heavy* propels *Starship* into orbit, a second *Starship* equipped as a tanker can refuel the spacecraft in space. This allows the spacecraft to leave Earth's

orbit fully fueled, increasing its range and payload capacity. Orbital refueling will be a game-changer for long-duration missions, reducing the need to carry all the necessary fuel at launch.

The Vision of a Self-Sustaining Mars Colony

At the core of SpaceX's mission is the vision of establishing a self-sustaining colony on Mars. Elon Musk has outlined a multi-phase plan to achieve this goal, beginning with initial exploration missions and eventually leading to the construction of a fully functioning city on Mars. The success of this plan hinges on the capabilities of *Starship* and the *Super Heavy* booster, as well as the development of technologies that enable humans to live and work on Mars.

1. **Initial Exploration and Research**: The first missions to Mars will focus on exploration and research, with the goal of understanding the planet's resources and environment. These missions will involve landing small crews on Mars to conduct scientific experiments, search for water ice, and test life support systems. *Starship* will serve as both the transportation vehicle and the habitat for these initial missions, providing shelter and supplies for the astronauts during their stay on the planet. These

missions will also scout potential sites for permanent bases and determine what technologies will be needed to support human life in the long term.

2. **Establishing a Permanent Base**: Once the initial research missions have identified suitable locations for human habitation and the necessary resources for survival, SpaceX's next step will be to establish a permanent base on Mars. This base will likely consist of multiple *Starship* vehicles converted into habitats, along with additional infrastructure for power generation, water extraction, and food production. The early base will depend on supplies from Earth, but the goal will be to gradually develop systems that allow the colony to produce its own resources.

3. **Resource Utilization and Independence**: A key element of SpaceX's plan is in-situ resource utilization (ISRU), which refers to the ability to harvest and use resources found on Mars. The planet has vast reserves of water ice, which can be melted for drinking water or split into hydrogen and oxygen for fuel and breathable air. Mars' atmosphere is primarily composed of carbon dioxide, which can be converted into methane (for fuel) and oxygen using

the Sabatier process. These capabilities are crucial for reducing the colony's dependence on resupply missions from Earth and ensuring its long-term sustainability.

4. **Building a Martian City**: Over time, SpaceX envisions the construction of a self-sustaining city on Mars, capable of supporting thousands or even millions of people. This city would include habitats, research labs, factories, farms, and other infrastructure necessary to maintain human life. One of the biggest challenges in building a Martian city will be protecting the inhabitants from the planet's harsh environment, including its low temperatures, thin atmosphere, and high levels of radiation. Advanced shielding, underground structures, and pressurized habitats will likely be used to address these issues.

5. **Terraforming Mars**: In the very long term, Elon Musk has proposed the idea of terraforming Mars—altering the planet's atmosphere and environment to make it more hospitable to human life. While this is a highly speculative and long-term goal, it would involve releasing greenhouse gases into the Martian

atmosphere to warm the planet, melt the polar ice caps, and create liquid water on the surface. Terraforming Mars is a monumental challenge that would likely take centuries to achieve, but it represents the ultimate vision of turning Mars into a second home for humanity.

How Starship and Super Heavy Enable Interplanetary Missions

The *Starship* and *Super Heavy* systems are essential to realizing the vision of interplanetary travel and colonization. Their unique design features and technological capabilities are what make missions to the Moon and Mars feasible.

1. **Reusability**: The fully reusable nature of both the *Super Heavy* booster and the *Starship* spacecraft is critical to lowering the cost of space travel. Traditional rockets are discarded after a single use, making each mission extremely expensive. By reusing rockets, SpaceX can dramatically reduce the cost of launching humans and cargo into space. This is essential for long-term missions to Mars, where multiple trips will be needed to transport people and supplies.

2. **Payload Capacity**: The sheer size and power of the *Super Heavy* booster allow it to carry large payloads into space. This is especially important for interplanetary missions, which require the transport of habitats, life support systems, scientific equipment, and other cargo needed for survival on another planet. *Starship* is designed to carry up to 100 metric tons of cargo, making it one of the most powerful spacecraft ever built.

3. **Orbital Refueling**: One of the key innovations that make *Starship* viable for long-duration missions to Mars is the ability to refuel in orbit. After being launched by the *Super Heavy* booster, *Starship* can be refueled by another *Starship* tanker in Earth's orbit, allowing it to leave Earth fully fueled and ready for the journey to Mars. This process extends *Starship*'s range and payload capacity, making it possible to transport larger loads to distant planets.

4. **Versatility**: The *Starship* spacecraft is designed to be highly versatile, capable of performing a wide range of missions. In addition to transporting people and cargo to Mars, *Starship* can be used for missions to the Moon, Earth's orbit, and even deep space

destinations like asteroids. This versatility makes *Starship* a multi-purpose vehicle that can support a wide range of scientific, commercial, and exploratory missions.

5. **Landing on Mars and Returning to Earth**: One of the biggest challenges of interplanetary travel is landing on and taking off from another planet. Mars' thin atmosphere makes it difficult to use parachutes or other traditional landing systems. *Starship* solves this problem by using a powered descent, where its Raptor engines are reignited to slow the spacecraft during its descent and guide it to a soft landing. Once on Mars, *Starship* can be refueled using locally sourced methane, allowing it to take off again and return to Earth.

The Road Ahead: Challenges and Opportunities

While *Starship* and *Super Heavy* represent a significant leap forward in space technology, there are still many challenges that must be overcome before regular interplanetary missions become a reality. These challenges include developing reliable life support systems, addressing the health risks posed by radiation, and finding ways to generate food, water, and energy on Mars.

However, the opportunities presented by *Starship* are vast. If successful, *Starship* will not only enable missions to Mars but also revolutionize space travel more broadly. It could be used for missions to the outer planets, asteroid mining, space tourism, and even the construction of space-based infrastructure such as space stations and habitats.

Additionally, *Starship* could play a key role in scientific research. By making space travel more affordable and accessible, *Starship* could enable more frequent missions to study distant planets, moons, and asteroids. This would increase our understanding of the solar system and potentially unlock new resources that could be used to support life on Earth or in space.

A Future Beyond Earth

SpaceX's *Starship* and *Super Heavy* represent the culmination of decades of innovation and the realization of Elon Musk's vision for making humanity a multi-planetary species. These spacecraft are designed not just to explore space but to enable humans to live and thrive on other worlds. From missions to the Moon as part of NASA's Artemis program to the long-term goal of establishing a city on Mars, *Starship* is at the forefront of a new era of space exploration.

While many challenges remain, the development of reusable, powerful spacecraft like *Starship* is bringing us closer to a future where interplanetary travel is routine, and the dream of living on Mars becomes a reality. SpaceX's vision is not just about exploring the cosmos—it's about ensuring the survival and expansion of human life beyond Earth. As we look to the stars, *Starship* stands as the vehicle that will take us there.

CHAPTER 7
Environmental and Regulatory Challenges

SpaceX's *Starship* development represents a revolutionary leap in space exploration, designed to make interplanetary travel feasible and affordable. However, the path to bringing *Starship* to life has not been without significant challenges, especially when it comes to environmental and regulatory hurdles. The construction and testing of *Starship*—particularly at SpaceX's Starbase in Boca Chica, Texas—have prompted concerns regarding environmental impacts, while regulatory approvals from agencies like the Federal Aviation Administration (FAA) have been critical in shaping the timeline of the project.

This chapter explores the various environmental and regulatory challenges SpaceX has faced throughout *Starship*'s development, how these hurdles affected the program's progress, and the complex balance between technological innovation and regulatory compliance.

The Environmental Impact of Space Launches

Space launches, by their nature, have significant environmental impacts. The construction and testing of rockets require vast amounts of energy, materials, and land, while the actual launches produce emissions and noise pollution that can affect surrounding ecosystems and communities. As SpaceX embarked on the development of the *Starship* system—a rocket far larger and more powerful than any of its predecessors—the environmental consequences of these activities became a central issue.

1. **Emissions and Pollution**: One of the primary environmental concerns associated with space launches is the emission of greenhouse gases and pollutants into the atmosphere. Rockets typically burn large amounts of fuel, producing carbon dioxide (CO_2), water vapor, and other emissions that can contribute to global warming. Additionally, solid rocket boosters (though not used by *Starship*) release chlorine gas, which can harm the ozone layer.

The *Starship* system, powered by SpaceX's Raptor engines, uses liquid methane and liquid oxygen as propellants. While methane burns more cleanly than traditional rocket fuels like kerosene or RP-1, the combustion process still produces carbon dioxide. Methane itself is a potent greenhouse gas, although its use in rocket propulsion is considered more environmentally friendly compared to other fuel options. Nevertheless, large-scale production and launches of *Starship* still raise concerns about the cumulative effect of emissions, especially if the goal is to launch frequently for missions to the Moon, Mars, and beyond.

2. **Noise Pollution**: Space launches generate tremendous noise, which can disrupt wildlife and disturb nearby communities. The *Super Heavy* booster, with its 33 Raptor engines, produces a level of thrust far greater than any rocket previously built. This results in extreme noise levels during launches, which can impact the habitats of birds and other wildlife, as well as the human populations living near SpaceX's launch facilities.

3. **Land Use and Habitat Destruction**: SpaceX's Starbase facility in Boca Chica, Texas, is situated near sensitive environmental areas, including wetlands, coastal habitats, and wildlife refuges. The development of the facility, along with the expansion needed to support the *Starship* project, has raised concerns about habitat destruction and the impact on

local wildlife, particularly endangered species such as sea turtles and migratory birds. The region around Boca Chica is home to a diverse array of wildlife, and the construction activities required to build launchpads, test facilities, and related infrastructure have caused disruption to these ecosystems.

4. **Marine Impact**: Launches from coastal areas like Boca Chica often involve rockets flying over the ocean, where the discarded stages of rockets and other debris can fall into the water. This debris can pose a hazard to marine life, pollute the water, and disrupt local fisheries. While SpaceX's focus on reusability (such as recovering the *Super Heavy* booster using the "chopsticks" mechanism) mitigates some of this impact, there is still concern about the environmental effects of ocean-based operations.

Regulatory Oversight: The Role of the FAA

The Federal Aviation Administration (FAA) is the primary regulatory body responsible for overseeing commercial spaceflight activities in the United States. The FAA's Office of Commercial Space Transportation (AST) is tasked with ensuring that spaceflight operations, such as those conducted by SpaceX, meet safety, environmental, and operational standards. The FAA has played a central role in regulating the development and testing of *Starship*, with its approvals determining when and how SpaceX can conduct launches.

1. **Licensing and Approval Process**: For each test flight or orbital launch of *Starship*, SpaceX must obtain a launch license from the FAA. These licenses are not granted automatically; they require a detailed review of the proposed launch's safety, environmental impact, and potential effects on nearby communities. The FAA's review process often involves collaboration with other agencies, such as the Environmental Protection Agency (EPA) and the U.S. Fish and Wildlife Service (FWS), to assess the environmental impact of rocket launches on wildlife, ecosystems, and public health.

2. **Environmental Assessments**: One of the key aspects of the FAA's regulatory role is conducting environmental assessments (EAs) or environmental impact statements (EISs) for proposed spaceflight activities. These assessments evaluate the potential effects of rocket launches, land use, and other activities on the surrounding environment. In the case of SpaceX's Starbase facility in Boca Chica,

environmental assessments have been critical in determining the viability of continued operations and expansions.

For example, in 2021, the FAA conducted an environmental review of SpaceX's plans to launch orbital *Starship* missions from Boca Chica. This review examined the potential impacts on wildlife, air and water quality, and the local community. Based on the findings of the assessment, the FAA could impose conditions on SpaceX's operations, such as limiting the number of launches, requiring noise abatement measures, or mandating additional conservation efforts to protect wildlife.

3. **Public and Legal Challenges**: SpaceX's rapid development timeline and high-profile testing campaigns have attracted significant public attention, leading to both support and opposition. While many view SpaceX's work as critical to advancing human space exploration, others have raised concerns about the environmental and regulatory implications of the company's activities. Several environmental organizations and local residents have filed legal challenges against SpaceX and the FAA, arguing that the agency's

environmental assessments have not adequately addressed the long-term impacts of rocket launches on the local environment.

In one high-profile case, environmental groups challenged the FAA's decision to grant SpaceX a launch license for *Starship* flights from Boca Chica, arguing that the agency failed to consider the cumulative environmental effects of the project. These legal challenges can delay SpaceX's plans, as the company must wait for the legal process to resolve before proceeding with certain operations.

Delays and Impact on Starship's Timeline

Environmental and regulatory challenges have had a significant impact on the timeline of *Starship*'s development. While SpaceX is known for its aggressive timelines and rapid testing, the need to comply with regulatory requirements and address environmental concerns has sometimes slowed progress. Several key delays can be attributed to these challenges:

1. **Orbital Launch Delays**: One of the most significant delays occurred in 2021 when SpaceX sought approval for the first orbital test flight of *Starship* from Boca Chica. The FAA's environmental review process, coupled with legal challenges from

environmental groups, delayed the issuance of a launch license. SpaceX had initially hoped to conduct the orbital flight by mid-2021, but the review process stretched into 2022, postponing the test flight by several months.

2. **Expansion of Starbase Facility**: SpaceX's plans to expand its Starbase facility in Boca Chica to support larger-scale operations—including the construction of new launchpads, integration towers, and infrastructure for *Starship* missions—faced regulatory scrutiny due to the potential environmental impact on the surrounding wetlands and wildlife refuges. The expansion plans required additional environmental assessments and consultations with federal agencies like the U.S. Fish and Wildlife Service, which added to the timeline.

3. **Mitigation Measures**: In response to concerns raised by regulatory agencies and environmental groups, SpaceX has implemented several mitigation measures to reduce the environmental impact of its operations. These measures include limiting the number of launches per year, scheduling launches to avoid sensitive times for wildlife (such as nesting

seasons), and working to minimize noise and light pollution. While these measures help protect the environment, they also impose operational constraints that can affect SpaceX's ability to meet its ambitious testing and launch schedules.

Balancing Innovation and Environmental Responsibility

SpaceX's drive to innovate and push the boundaries of space exploration has often been at odds with the need to balance environmental responsibility and regulatory compliance. While SpaceX's leadership has been vocal about the importance of advancing space technology to benefit humanity, the company has also faced criticism for its perceived disregard for environmental concerns in the rush to develop *Starship*.

1. **SpaceX's Environmental Commitments**: In response to growing concerns, SpaceX has taken steps to demonstrate its commitment to environmental sustainability. For example, the company has pledged to work with local environmental organizations to protect wildlife near the Starbase facility and has implemented measures to reduce the environmental footprint of its operations. Additionally, SpaceX's focus on

developing fully reusable rockets, such as *Super Heavy* and *Starship*, is seen as a key step toward reducing the environmental impact of space launches. By reusing rockets rather than discarding them after a single use, SpaceX aims to minimize waste and reduce the need for large-scale rocket production, which has its own environmental costs.

2. **Technological Solutions**: SpaceX is also exploring technological solutions to mitigate the environmental effects of its operations. One example is the development of cleaner propulsion technologies. The Raptor engines used in the *Super Heavy* and *Starship* systems are designed to burn methane and oxygen, which produce fewer pollutants compared to traditional rocket fuels. Additionally, methane can potentially be sourced from renewable or Martian sources, further reducing the environmental impact of rocket operations over the long term.

3. **Engagement with Regulatory Agencies**: Despite engagement with regulatory agencies and stakeholders has been essential for SpaceX to maintain its development timeline and meet its environmental responsibilities. By working closely

with the FAA, the U.S. Fish and Wildlife Service, and other agencies, SpaceX has been able to navigate the complex regulatory landscape surrounding spaceflight activities. The company has participated in public hearings, environmental reviews, and consultations to ensure that its operations comply with federal and state laws.

4. **Public Relations and Transparency**: SpaceX has also worked to improve its transparency regarding environmental and regulatory issues. The company has provided updates on its environmental initiatives, hosted public outreach events, and engaged with the local community to address concerns about noise, pollution, and habitat disruption. These efforts have helped build public trust and foster a more collaborative relationship with regulators, environmental groups, and residents in the areas surrounding SpaceX's launch sites.

SpaceX's Future Challenges in Environmental and Regulatory Compliance

As SpaceX continues to scale up its operations and move toward its goal of frequent, reusable space travel, the company will face ongoing environmental and regulatory challenges. The development of new launch sites, the construction of additional infrastructure, and the increasing number of launches will require careful planning to balance the need for innovation with the responsibility to protect the environment.

1. **Global Expansion**: As SpaceX looks to expand its operations beyond Boca Chica, the company will need to navigate a new set of environmental and regulatory requirements in other regions. SpaceX has already established launch facilities at Cape Canaveral in Florida and Vandenberg Air Force Base in California, but as the demand for space launches grows, the company may explore new launch sites, both domestically and internationally. Each new site will require environmental assessments, land-use approvals, and consultations with local governments and regulatory agencies.

2. **Climate Change and Carbon Emissions**: While *Starship* and *Super Heavy* represent significant advancements in rocket reusability, the issue of carbon emissions remains a concern for the broader space industry. As climate change becomes an increasingly urgent global issue, regulatory agencies

may impose stricter limits on the emissions generated by space launches. SpaceX will need to continue exploring ways to reduce the carbon footprint of its operations, possibly by investing in carbon capture technologies, renewable energy sources for its facilities, and more efficient propulsion systems.

3. **Space Debris and Orbital Pollution**: Another emerging regulatory concern is the issue of space debris. As SpaceX launches more missions, especially with its *Starlink* satellite constellation, the company will need to address the growing problem of orbital congestion and debris. Regulatory agencies, including the FAA and the Federal Communications Commission (FCC), are likely to implement stricter guidelines for satellite deployment, deorbiting procedures, and space traffic management to prevent collisions and minimize the accumulation of debris in low Earth orbit. SpaceX will need to work with international regulatory bodies to ensure that its operations comply with emerging standards for space sustainability.

Navigating the Intersection of Innovation and Regulation

The development of SpaceX's *Starship* system has not only redefined what is technologically possible in space exploration but has also highlighted the complex balance between innovation and regulation. The company's efforts to bring *Starship* to life have faced significant environmental and regulatory hurdles, from concerns about habitat destruction and emissions to the need for extensive oversight by the FAA and other federal agencies.

Despite these challenges, SpaceX has demonstrated a willingness to engage with regulators, adapt its operations to minimize environmental impacts, and comply with the laws that govern spaceflight activities. The company's focus on reusability, cleaner propulsion technologies, and rapid turnaround times is paving the way for a future where space travel is not only more accessible but also more sustainable.

As SpaceX continues its journey toward Mars and beyond, the company's success will depend not only on its engineering prowess but also on its ability to navigate the regulatory and environmental landscape. The path to interplanetary travel may be long and fraught with challenges, but with careful planning, collaboration, and a commitment to responsible innovation, SpaceX is well-positioned to lead humanity into the next era of space exploration.

CHAPTER 8

The Future of Space Travel: Commercial and Scientific Impacts

The successful development of SpaceX's *Starship* system is set to have far-reaching consequences not only for space exploration but for the broader commercial space industry. By making space travel more accessible and affordable, *Starship* has the potential to revolutionize satellite launches, cargo transport, space tourism, and even scientific research. The *Starship* system's fully reusable design, coupled with its unparalleled payload capacity, places it in a position to reshape the way we think about space travel—turning it from a rare and costly endeavor into an everyday possibility for businesses, governments, and private individuals alike.

This chapter will explore the potential commercial and scientific impacts of *Starship*, analyzing how SpaceX's innovations could drive new industries, enhance scientific exploration, and usher in a new era of human activity in space.

Revolutionizing Commercial Space Travel

SpaceX's *Starship* system represents a paradigm shift in space travel, particularly when it comes to cost and reusability. Traditional space travel has been prohibitively expensive, with the cost of launching payloads into orbit often reaching tens or even hundreds of millions of dollars. Rockets were typically single-use, meaning they could only be flown once before being discarded, which added to the overall expense of each mission.

SpaceX, through the development of reusable rockets such as *Falcon 9* and *Starship*, has dramatically reduced the cost of launching payloads into space. This reduction in cost has profound implications for the commercial space industry, enabling more companies, governments, and research institutions to conduct space missions that were previously out of reach due to financial constraints.

Lowering Launch Costs

The development of reusable rockets like *Starship* has the potential to lower the cost per kilogram of launching cargo into space to an unprecedented degree. Elon Musk has stated that the marginal cost of a single *Starship* launch could eventually drop to as low as $2 million, a fraction of the cost of traditional rockets. This cost reduction is primarily driven by reusability, as the ability to recover and reuse both the *Super Heavy* booster and the *Starship* spacecraft means that new rockets do not need to be built for each launch.

For commercial entities, this reduction in cost opens up a range of new opportunities. Telecommunications companies, for example, can launch larger and more sophisticated satellite constellations at a fraction of the previous cost, improving global communication networks and expanding internet access to underserved regions. Similarly, research institutions can launch more experiments and scientific instruments into space, furthering our understanding of Earth, space weather, and the broader universe.

Scaling Satellite Deployment

One of the most immediate commercial impacts of *Starship* will be in the satellite industry. The global satellite market, which is critical for communications, weather forecasting, navigation, and Earth observation, stands to benefit significantly from the capabilities of *Starship*. The spacecraft's payload capacity—up to 100 metric tons—will allow it to deploy large numbers of satellites in a single launch, reducing the overall cost per satellite and enabling the deployment of mega-constellations like SpaceX's *Starlink* network.

SpaceX's *Starlink* is a prime example of how *Starship* could revolutionize satellite deployment. *Starlink* aims to provide global high-speed internet coverage using a constellation of thousands of small satellites. With *Falcon 9*, SpaceX has already launched thousands of *Starlink* satellites, but *Starship* is expected to greatly accelerate the process by enabling the deployment of larger batches of satellites in a single launch.

Other satellite companies will also benefit from *Starship*'s capabilities. Companies like One Web, Amazon's Project Kuiper, and others looking to deploy large satellite constellations will be able to take advantage of SpaceX's lower launch costs and higher payload capacities, making global satellite networks more viable and affordable.

Enabling Space Tourism

Space tourism has long been a dream for those with the resources and desire to travel beyond Earth's atmosphere. However, the high cost of spaceflight has historically limited space tourism to a small number of ultra-wealthy individuals and private missions. The development of *Starship* could change this dynamic by making space travel more accessible to a broader audience.

SpaceX has already signaled its interest in space tourism by signing agreements with several high-profile individuals and organizations for private *Starship* flights. For example, Japanese entrepreneur Yusaku Maezawa has contracted with SpaceX for a private mission around the Moon, dubbed the "Dear Moon" project. This mission, set to carry Maezawa and a group of artists and cultural figures, will be one of the first human missions on *Starship* and a major milestone in the commercial space tourism industry.

As *Starship* flights become more frequent and reliable, the cost of space tourism is expected to decrease, allowing a larger number of private individuals to experience space travel. *Starship*'s large payload capacity means it can carry multiple passengers on each mission, further reducing the cost per person. Space tourism companies, such as Blue Origin and Virgin Galactic, are already working to provide suborbital spaceflights for private customers, but *Starship*'s orbital and deep-space capabilities could set it apart by offering more immersive and extended space experiences.

In the future, *Starship* could also be used for private missions to the Moon, Mars, and other deep-space destinations. As

SpaceX continues to refine the spacecraft's technology and reduce launch costs, space tourism could evolve into a mainstream industry, offering once-in-a-lifetime experiences to a growing number of people.

Cargo Transport and Space Logistics

In addition to satellite deployment and space tourism, *Starship* has the potential to revolutionize cargo transport and space logistics. As humanity's activities in space expand, so too will the need for reliable transportation of cargo, supplies, and resources. Space agencies, private companies, and future space colonies will all require a steady flow of materials to sustain operations in space, and *Starship* is uniquely positioned to meet this demand.

Resupplying the International Space Station (ISS)

One of the first commercial applications of *Starship* could be its use as a resupply vehicle for the International Space Station (ISS) and other space stations that may be developed in the future. Currently, NASA relies on a combination of government and commercial spacecraft—such as SpaceX's *Dragon* and Northrop Grumman's *Cygnus*—to transport cargo to the ISS. However, *Starship*'s massive payload capacity would allow it to carry far more supplies in a single trip, making resupply missions more efficient and cost-effective.

The same principles apply to future space stations, such as those being developed by private companies like Axiom Space or Blue Origin's Orbital Reef. As the commercial space industry grows, *Starship* could become the go-to vehicle for transporting cargo, crew, and equipment to orbital habitats and research facilities.

Supporting Lunar and Martian Colonies

Looking further ahead, *Starship* could play a central role in establishing and supporting lunar and Martian colonies. SpaceX's long-term vision is to enable humanity to become a multi-planetary species, with permanent settlements on the Moon and Mars. For these colonies to be viable, regular cargo missions will be necessary to deliver food, water, equipment, and building materials.

The *Starship* system's ability to carry large payloads across long distances makes it ideal for supporting these missions. On the Moon, *Starship* could be used to deliver habitats, rovers, and scientific instruments to the surface, while also serving as a crewed lander as part of NASA's Artemis program. For Mars, *Starship* would be critical for transporting the supplies needed to build a self-sustaining colony, including solar panels, life support systems, and manufacturing equipment.

Interplanetary Supply Chains

In the long term, *Starship* could also enable the development of interplanetary supply chains, where goods and resources are transported between Earth, the Moon, Mars, and potentially other destinations in the solar system. This could include everything from raw materials mined from asteroids to manufactured goods produced in space. The establishment of interplanetary supply chains would mark a new phase in human civilization, where space becomes not just a place for exploration, but a frontier for industry and commerce.

Scientific Impacts of Starship

While the commercial applications of *Starship* are vast, the spacecraft also has the potential to drive significant advancements in scientific research. By making space more accessible and affordable, *Starship* opens up new opportunities for scientific exploration, both in Earth's orbit and across the solar system.

Expanding Access to Space for Researchers

One of the most immediate scientific impacts of *Starship* is its potential to democratize access to space for researchers.

Currently, space research is limited by the high cost of launching scientific instruments and experiments into orbit. While space agencies like NASA, the European Space Agency (ESA), and private companies like SpaceX provide opportunities for research missions, the cost constraints limit the number of projects that can be funded.

By dramatically lowering the cost of launching payloads, *Starship* could allow universities, research institutions, and even private companies to conduct experiments in space that were previously out of reach. This could lead to breakthroughs in fields such as microgravity research, materials science, and space medicine. Space agencies could also use *Starship* to deploy more sophisticated scientific instruments, such as space telescopes or probes, to study distant planets, stars, and galaxies.

Lunar and Martian Exploration

One of the most exciting scientific opportunities presented by *Starship* is the potential for exploring the Moon and Mars in greater detail. While robotic missions have provided valuable data about these celestial bodies, human exploration offers the opportunity to conduct more complex and nuanced research.

Lunar and Martian Exploration (Continued)

One of the most exciting scientific opportunities presented by *Starship* is the potential for exploring the Moon and Mars in greater detail. While robotic missions have provided valuable data about these celestial bodies, human exploration offers the opportunity to conduct more complex and nuanced research.

For the Moon, *Starship* could be used to transport scientific instruments, rovers, and crew to regions that have yet to be explored in detail, such as the lunar poles. These areas are of particular interest because they contain water ice deposits, which could be used to support future lunar bases. Human explorers aboard *Starship* could also collect geological samples, conduct in-situ resource utilization (ISRU) experiments, and establish permanent outposts to serve as platforms for scientific research.

On Mars, *Starship*'s capabilities would allow for the first human expeditions to the surface, where astronauts could conduct detailed geological surveys, search for signs of past or present life, and test technologies needed for long-term habitation. The ability to return samples from Mars to Earth for detailed analysis would mark a major milestone in planetary science, helping researchers to better understand the planet's history and potential for supporting life.

Additionally, *Starship*'s ability to transport large amounts of equipment means that future Mars missions could establish more advanced laboratories and habitats, allowing for more extensive research and experimentation on the surface. This would be a critical step toward realizing SpaceX's goal of creating a self-sustaining colony on Mars, where scientific research and resource extraction could be carried out on a large scale.

Deep Space Exploration and Interplanetary Probes

Beyond the Moon and Mars, *Starship* has the potential to transform the way we explore the outer reaches of the solar system. The spacecraft's large payload capacity and ability to carry heavy scientific instruments make it an ideal platform for launching interplanetary probes and landers to destinations such as Jupiter's moons, Saturn's rings, or even asteroids in the Kuiper Belt.

With *Starship*, scientists could send more ambitious missions to explore these distant worlds, gathering data that could provide insights into the formation of the solar system and the potential for life beyond Earth. For example, a mission to Europa, one of Jupiter's moons, could involve landing a probe on the icy surface and drilling into the subsurface ocean to search for signs of microbial life. Such a mission would require substantial payloads, advanced scientific equipment, and the ability to transport large amounts of power and communications gear—capabilities that *Starship* can provide.

Similarly, *Starship* could be used to deploy deep-space telescopes or other scientific instruments that are too large or complex to launch with traditional rockets. By enabling the construction and deployment of more advanced space-based observatories, *Starship* could help scientists study distant galaxies, black holes, and other cosmic phenomena in greater detail than ever before.

Supporting International Collaboration

Starship is also poised to play a significant role in fostering international collaboration in space exploration. As the spacecraft's capabilities continue to evolve, it is likely that space agencies around the world, such as NASA, the European Space Agency (ESA), and the Japan Aerospace Exploration Agency (JAXA), will seek to partner with SpaceX for joint missions to the Moon, Mars, and beyond.

By providing a cost-effective and reliable platform for launching crewed and uncrewed missions, *Starship* can help reduce the barriers to entry for countries that may not have the resources to develop their own heavy-lift launch vehicles. This could lead to greater international cooperation in space exploration, with multiple countries contributing scientific instruments, technologies, and expertise to joint missions.

For example, NASA has already selected *Starship* as the lunar lander for its Artemis program, which aims to return humans to the Moon by 2025. As part of this program, *Starship* will carry astronauts from lunar orbit to the surface, where they will conduct scientific research and test technologies for future missions to Mars. This collaboration between SpaceX and NASA is just one example of how *Starship* could be used to support international efforts in space exploration.

Potential Commercial and Economic Impacts

While the scientific potential of *Starship* is immense, its commercial implications are equally profound. The development of reusable, cost-effective space transportation systems like *Starship* could unlock new industries and create significant economic opportunities.

1. **Space Mining**: One of the most talked-about future industries is space mining—the extraction of

valuable resources such as water, metals, and minerals from asteroids, the Moon, or other celestial bodies. *Starship* could play a central role in enabling this industry by providing the heavy-lift capabilities needed to transport mining equipment to space and return valuable resources to Earth. Asteroids, in particular, contain vast amounts of precious metals such as platinum and gold, which could be harvested for use in industrial applications on Earth.

2. **Space Manufacturing**: The development of space-based manufacturing facilities is another area where *Starship* could have a significant impact. By reducing the cost of transporting materials and equipment to space, *Starship* could enable companies to build factories in orbit or on the surface of the Moon, where unique manufacturing conditions—such as microgravity and a lack of atmospheric interference—could be used to produce advanced materials and products that are difficult or impossible to make on Earth.

3. **Space Habitats**: As space tourism and other commercial activities expand, the demand for space habitats and infrastructure will grow. Companies like

Axiom Space and Bigelow Aerospace are already developing commercial space stations that could serve as hotels, research labs, or manufacturing facilities. *Starship* could be used to transport modules, supplies, and personnel to these habitats, supporting the growth of a space-based economy.

4. **Telecommunications and Connectivity**: The deployment of large satellite constellations by companies like SpaceX (with *Starlink*) and Amazon (with Project Kuiper) will have a profound impact on global telecommunications. By providing affordable, high-speed internet access to remote and underserved regions, these constellations will bridge the digital divide and enable economic growth in areas that currently lack reliable connectivity. *Starship*'s ability to launch large numbers of satellites in a single mission will be key to the success of these initiatives.

The Future of Human Space Exploration

Ultimately, the greatest impact of *Starship* may be its role in enabling humanity to become a multi-planetary species. SpaceX's long-term vision is to establish self-sustaining colonies on the Moon and Mars, where humans can live, work, and thrive independent of Earth. This vision is rooted in the belief that expanding human civilization to other planets is essential for ensuring the long-term survival of our species.

1. **Colonizing the Moon**: The Moon will likely be the first destination for human colonization, serving as a testing ground for the technologies and systems needed to support long-term habitation in space. *Starship* could be used to transport crew, cargo, and infrastructure to the Moon, where permanent bases could be established to support scientific research, mining, and other activities. These bases could also serve as waypoints for missions to Mars and other destinations in the solar system.

2. **Colonizing Mars**: Mars, however, remains the ultimate goal for SpaceX and other space agencies. The Red Planet offers the potential for human settlement, thanks to its relative proximity to Earth and the presence of key resources such as water ice. Establishing a self-sustaining colony on Mars will be one of the most challenging and ambitious projects in human history, but *Starship* is designed with this goal in mind.

Axiom Space and Bigelow Aerospace are already developing commercial space stations that could serve as hotels, research labs, or manufacturing facilities. *Starship* could be used to transport modules, supplies, and personnel to these habitats, supporting the growth of a space-based economy.

4. **Telecommunications and Connectivity**: The deployment of large satellite constellations by companies like SpaceX (with *Starlink*) and Amazon (with Project Kuiper) will have a profound impact on global telecommunications. By providing affordable, high-speed internet access to remote and underserved regions, these constellations will bridge the digital divide and enable economic growth in areas that currently lack reliable connectivity. *Starship*'s ability to launch large numbers of satellites in a single mission will be key to the success of these initiatives.

The Future of Human Space Exploration

Ultimately, the greatest impact of *Starship* may be its role in enabling humanity to become a multi-planetary species. SpaceX's long-term vision is to establish self-sustaining colonies on the Moon and Mars, where humans can live, work, and thrive independent of Earth. This vision is rooted in the belief that expanding human civilization to other planets is essential for ensuring the long-term survival of our species.

1. **Colonizing the Moon**: The Moon will likely be the first destination for human colonization, serving as a testing ground for the technologies and systems needed to support long-term habitation in space. *Starship* could be used to transport crew, cargo, and infrastructure to the Moon, where permanent bases could be established to support scientific research, mining, and other activities. These bases could also serve as waypoints for missions to Mars and other destinations in the solar system.

2. **Colonizing Mars**: Mars, however, remains the ultimate goal for SpaceX and other space agencies. The Red Planet offers the potential for human settlement, thanks to its relative proximity to Earth and the presence of key resources such as water ice. Establishing a self-sustaining colony on Mars will be one of the most challenging and ambitious projects in human history, but *Starship* is designed with this goal in mind.

cargo to lunar bases, or enabling the first human colony on Mars, *Starship* is poised to usher in a new era of space exploration—one in which space becomes not just the final frontier, but a place where humanity can build a sustainable future.

3. **Terraforming Mars**: In the very long term, some scientists and visionaries—including Elon Musk—have proposed the idea of terraforming Mars, transforming it into a more Earth-like planet by releasing greenhouse gases into the atmosphere to warm the planet and create liquid water. While this concept remains speculative and would likely take centuries to achieve, *Starship* could play a key role in delivering the infrastructure and resources needed to begin the process.

Conclusion: A New Era in Space Exploration

SpaceX's *Starship* system represents a transformative moment in the history of space travel. By dramatically reducing the cost of access to space and increasing payload capacity, *Starship* has the potential to revolutionize industries ranging from satellite deployment to space tourism to scientific research. Its role in enabling human exploration of the Moon, Mars, and beyond cannot be overstated, as it opens up new frontiers for both commercial and scientific endeavors.

As *Starship* continues to undergo testing and development, the possibilities for its use will only expand. Whether it's launching mega-constellations of satellites, transporting

CHAPTER 9

What's Next for Starship?

SpaceX's *Starship* system represents a pivotal moment in the evolution of space travel. With its bold vision of making life multi-planetary, the project has already reached significant milestones, but much of its potential remains untapped. As *Starship* continues to develop, the next steps will be crucial in determining the future of human exploration beyond Earth. These steps include refining the spacecraft, testing its ability to refuel in orbit, and preparing for further manned missions to the Moon and Mars. In this chapter, we'll explore the upcoming challenges, milestones, and exciting possibilities that lie ahead for *Starship* and SpaceX.

Refueling Starship in Orbit: A Key Milestone

One of the most critical next steps in the *Starship* program is developing and demonstrating the ability to refuel the spacecraft in orbit. This is a fundamental component of SpaceX's long-term goal to make *Starship* capable of long-duration missions to the Moon, Mars, and beyond.

The concept of orbital refueling involves launching a second *Starship* into space specifically designed as a tanker. This tanker would rendezvous with the crewed or cargo-carrying *Starship* in low-Earth orbit (LEO) and transfer propellant to it. This process allows *Starship* to carry less fuel during its initial launch, maximizing the amount of cargo or crew it can transport. Once refueled in space, *Starship* would have the necessary propellant to travel long distances, such as to Mars or the outer planets.

Why Refueling Matters

Orbital refueling is essential for missions beyond low-Earth orbit due to the significant amount of fuel required to escape Earth's gravity and travel deep into space. Without refueling, a fully loaded *Starship* would not have enough fuel to complete a round-trip mission to Mars and return to Earth. The refueling process enables SpaceX to overcome the constraints of payload and distance by topping off fuel tanks after leaving Earth, allowing for longer, more ambitious missions.

Technical Challenges

While the idea of refueling in space is straightforward in theory, it presents several technical challenges:

1. **Cryogenic Propellant Storage**: *Starship* uses liquid methane and liquid oxygen as propellants, both of

which must be stored at extremely low temperatures. Ensuring that these cryogenic fuels remain stable during the time they spend in space and can be transferred between two spacecraft without evaporating presents a significant engineering challenge. Advanced insulation and thermal management systems are required to prevent boil-off, the process by which cryogenic liquids turn into gas and are lost to space.

2. **Docking in Orbit**: Orbital refueling will require *Starship* to dock with a tanker spacecraft in the microgravity environment of space. Docking two large spacecraft, each potentially filled with volatile fuel, requires precision and safety. SpaceX has already demonstrated some level of docking expertise through the *Dragon* spacecraft's missions to the International Space Station (ISS), but docking two large vehicles like *Starship* will be more complex.

3. **Automated Fuel Transfer**: The process of transferring fuel in space must be fully automated and carefully controlled. SpaceX will need to develop specialized hardware, including pumps,

valves, and pipelines, that can safely transfer cryogenic fuel from one spacecraft to another in zero gravity. The system must prevent leaks or spills, which could be dangerous in the vacuum of space.

Orbital Refueling Tests and Timeline

SpaceX has not yet demonstrated orbital refueling, but it is an essential milestone that will need to be achieved before manned missions to Mars can become a reality. Musk has stated that multiple *Starship* refueling missions may be required to fully fuel a single *Starship* for a journey to Mars. This makes the efficiency and reliability of the refueling process a critical factor in the feasibility of interplanetary travel.

Testing the refueling process is expected to occur in the near future, potentially after *Starship* completes its first orbital flight and SpaceX gathers more data on its performance in space. Once successful, orbital refueling will allow SpaceX to take the next steps toward more distant missions, such as those to Mars and even deep-space destinations like Jupiter's moons.

Further Manned Missions to the Moon

While Mars is the ultimate goal, SpaceX's next significant human spaceflight milestone will involve further manned missions to the Moon. NASA's Artemis program has selected *Starship* as the lander for the Artemis III mission, which will carry astronauts to the lunar surface by 2025. This mission will mark the first time humans have set foot on the Moon since the Apollo program and will be a critical test of *Starship*'s capabilities for deep-space exploration.

The Artemis Program

NASA's Artemis program aims to return humans to the Moon as part of a long-term effort to establish a sustainable presence on the lunar surface. Unlike the short-duration missions of Apollo, Artemis aims to build a lunar base where astronauts can live and work for extended periods. This base will serve as a platform for scientific research, testing new technologies, and preparing for future missions to Mars.

SpaceX's *Starship* was selected as NASA's Human Landing System (HLS) for the Artemis III mission, a major vote of confidence in the spacecraft's design and potential. The mission will involve using NASA's Space Launch System (SLS) rocket to send astronauts to lunar orbit, where they will transfer to *Starship* for the descent to the lunar surface.

The Role of Starship in Lunar Exploration

Starship is uniquely suited to lunar exploration due to its large payload capacity and reusability. Unlike the Apollo lunar module, which was only capable of carrying a small crew and limited supplies, *Starship* can transport large amounts of cargo, habitats, and scientific equipment to the lunar surface. This capability will be essential for building a sustainable lunar base, as it allows astronauts to bring everything they need for long-duration missions.

Starship's reusability is another critical advantage. After landing on the Moon, the spacecraft can be refueled and relaunched for additional missions, reducing the overall cost of lunar exploration. This is in stark contrast to the expendable lunar modules of the Apollo era, which could only be used once before being discarded.

Lunar Surface Operations

Once *Starship* lands on the Moon, it will serve as both a habitat and a launch platform for future missions. The spacecraft's spacious interior provides ample room for astronauts to live and work, while its powerful engines enable it to take off from the Moon's surface and return to orbit. This will allow astronauts to conduct extended scientific research, exploring the lunar surface and collecting samples over a longer period.

NASA and SpaceX also plan to use *Starship* for a variety of lunar surface activities, including establishing infrastructure such as power systems, habitats, and transportation networks. These activities will be crucial for laying the groundwork for a permanent human presence on the Moon.

The Path to Mars: What's Next?

While *Starship*'s lunar missions will be a significant achievement in their own right, Mars remains the ultimate destination for SpaceX and its founder, Elon Musk. SpaceX has made it clear that the primary goal of the *Starship* program is to enable human missions to Mars, with the long-term objective of establishing a self-sustaining colony on the Red Planet.

Key Challenges of Mars Missions

Mars presents a host of unique challenges that make it a far more difficult destination than the Moon. While the Moon is relatively close to Earth and has been explored extensively, Mars is a distant world with a harsh environment, thin atmosphere, and extreme temperatures. Successful missions to Mars will require overcoming several major obstacles:

1. **Distance and Travel Time**: A trip to Mars takes significantly longer than a trip to the Moon. Depending on the alignment of the planets, the journey to Mars can take anywhere from six to nine months. This means astronauts will need to spend an extended period in space, requiring advanced life support systems, radiation protection, and psychological support.

2. **Landing on Mars**: Mars' thin atmosphere makes landing large spacecraft on its surface extremely challenging. The atmosphere is too thin to provide significant aerodynamic drag, but thick enough to generate heat and stress during reentry. SpaceX plans to use *Starship*'s Raptor engines for a powered descent, similar to the way the *Falcon 9* booster lands on Earth, but at much higher velocities and with less margin for error.

3. **Sustaining Life on Mars**: Once on the surface of Mars, astronauts will face numerous challenges in sustaining life. The planet's atmosphere is primarily composed of carbon dioxide, with little oxygen or water available. SpaceX will need to develop advanced life support systems, habitats, and resource

utilization technologies to produce air, water, and food from the Martian environment.

4. **Returning to Earth**: Unlike lunar missions, where astronauts can return to Earth relatively quickly, a mission to Mars will require astronauts to stay on the planet for an extended period, waiting for the next favorable launch window. SpaceX plans to use in-situ resource utilization (ISRU) to produce methane and oxygen fuel from Mars' atmosphere, allowing *Starship* to refuel on the surface and return to Earth.

Milestones Leading to the First Mars Mission

Before humans can set foot on Mars, SpaceX will need to achieve several key milestones. These include:

1. **Orbital Refueling**: As discussed earlier, orbital refueling is essential for enabling *Starship* to carry enough fuel for a round-trip mission to Mars. Demonstrating this capability will be one of the most critical milestones leading up to a manned mission.

2. **Unmanned Mars Missions**: Before sending humans to Mars, SpaceX plans to conduct several unmanned missions to test *Starship*'Before sending humans to Mars, SpaceX plans to conduct several unmanned

missions to test *Starship*'s capabilities. These missions will likely involve landing *Starship* on the Martian surface, testing key systems such as life support, in-situ resource utilization (ISRU), and surface mobility. These unmanned missions will provide critical data on how the spacecraft performs in the harsh Martian environment and help SpaceX refine its systems before attempting a crewed mission.

3. **In-Situ Resource Utilization (ISRU)**: One of the key technologies that will enable sustainable Mars missions is ISRU, which involves using the resources available on Mars to produce fuel, oxygen, and water. SpaceX plans to use the Sabatier process to convert carbon dioxide from Mars' atmosphere into methane and oxygen, which can be used as fuel for *Starship's* return trip to Earth. Demonstrating ISRU on Mars will be a major milestone in the path to sustainable human exploration of the planet.

4. **Building a Base on Mars**: Once *Starship* has successfully landed humans on Mars, the next step will be to establish a permanent base on the planet. This base will serve as a hub for scientific research,

resource extraction, and the development of technologies needed to sustain human life on Mars. The base will likely start as a small outpost, with astronauts living and working inside *Starship* or inflatable habitats, before expanding into a larger, more self-sufficient colony.

Long-Term Goals: A Self-Sustaining Colony on Mars

Elon Musk's ultimate vision for *Starship* is to enable the establishment of a self-sustaining human colony on Mars. This would be the culmination of SpaceX's efforts to make life multi-planetary, ensuring the survival of the human species in the event of a catastrophe on Earth.

1. **Terraforming Mars**: In the very long term, Musk has proposed the idea of terraforming Mars—altering the planet's atmosphere and climate to make it more hospitable for human life. While this concept is highly speculative and would likely take centuries to achieve, it represents the ultimate goal of creating a second home for humanity on another planet. Terraforming would involve releasing greenhouse gases to warm the planet and create a thicker atmosphere, which could eventually support liquid water and vegetation.

2. **Expanding Human Civilization**: As the colony on Mars grows, it could serve as a stepping stone for further exploration of the solar system. Mars could become a hub for missions to the asteroid belt, the moons of Jupiter and Saturn, and beyond. By establishing a permanent presence on Mars, humanity would take its first steps toward becoming an interplanetary civilization.

What's Next for Starship: Upcoming Milestones

As *Starship* continues to evolve, there are several key milestones and missions to watch for in the coming years. These include:

1. **First Orbital Flight of Starship**: SpaceX has conducted multiple high-altitude test flights of *Starship* prototypes, but the next major milestone will be the first orbital flight of a fully integrated *Starship* and *Super Heavy* booster. This flight will demonstrate *Starship's* ability to reach orbit and re-enter Earth's atmosphere, paving the way for future missions to the Moon and Mars.

2. **Orbital Refueling Demonstration**: As discussed earlier, demonstrating the ability to refuel *Starship* in orbit will be critical for enabling long-duration

missions to Mars. SpaceX plans to conduct tests of orbital refueling in the coming years, using tanker versions of *Starship* to transfer fuel in space.

3. **NASA's Artemis III Mission**: One of the most exciting milestones for *Starship* will be its role in NASA's Artemis III mission, which aims to land humans on the Moon by 2025. This mission will be a major test of *Starship's* capabilities and will mark the first time humans have returned to the Moon in over 50 years.

4. **Unmanned Missions to Mars**: Before sending humans to Mars, SpaceX plans to conduct several unmanned missions to test *Starship's* ability to land on the Martian surface and perform key operations such as ISRU. These missions will provide valuable data on how *Starship* performs in the Martian environment and help SpaceX prepare for crewed missions.

5. **First Crewed Mission to Mars**: The ultimate milestone for *Starship* will be the first crewed mission to Mars. While the exact timeline for this mission is still uncertain, Musk has stated that he hopes to send humans to Mars by the late 2020s or

early 2030s. This mission will mark the beginning of humanity's journey to become a multi-planetary species.

The Future of Space Travel with Starship

SpaceX's *Starship* system represents the future of space travel. With its fully reusable design, large payload capacity, and ability to perform long-duration missions, *Starship* has the potential to revolutionize not only exploration but also the commercial space industry, satellite launches, cargo transport, and space tourism.

As SpaceX continues to refine *Starship* and overcome technical challenges such as orbital refueling and landing on Mars, the spacecraft will play a central role in humanity's efforts to explore the solar system and establish permanent colonies on other planets. The upcoming milestones, from the first orbital flight to the first manned mission to Mars, will be critical in shaping the future of space exploration and determining whether SpaceX can achieve its ambitious goal of making life multi-planetary.

In the coming years, *Starship* will take humanity to new frontiers, opening the door to a future where space travel is not just the domain of government space agencies, but a thriving industry that touches every aspect of human life.

Whether it's building a lunar base, sending astronauts to Mars, or enabling space tourism for millions, *Starship* is poised to transform the way we think about our place in the universe.

CONCLUSION

The Vision and Reality of SpaceX's Starship System

SpaceX's *Starship* system, culminating in the ability of the *Starship Engine Core* to return to base using the innovative "chopsticks" mechanism, represents a monumental leap in both technological innovation and human ambition. What started as Elon Musk's dream of making humanity a multi-planetary species has transformed into one of the most ambitious engineering projects in modern history. Through relentless testing, iteration, and groundbreaking advancements, SpaceX has set the stage for a new era of space exploration and commercial travel.

This book has explored the various stages and technical aspects that led to the success of SpaceX's *Starship* system, focusing on the reusability of the *Super Heavy* booster and the precision landings enabled by the *chopsticks* mechanism. Alongside this, we've examined the environmental and regulatory challenges SpaceX has had to overcome, the future implications of the *Starship* system for both commercial and scientific ventures, and the next steps that will take SpaceX from Earth to Mars.

The Significance of Reusability

At the core of SpaceX's vision is the concept of reusability. Traditional space travel has been hindered by the prohibitively high costs associated with building new rockets for each mission, making exploration an expensive and rare endeavor. By developing reusable rocket technology, SpaceX aims to dramatically reduce the cost of space travel, making it as routine and affordable as air travel.

The ability of the *Super Heavy* booster to return to its launch site using the *Mechazilla* tower and its chopsticks mechanism is a pivotal moment in this journey toward full reusability. Not only does this system eliminate the need for costly recovery operations at sea, but it also minimizes wear and tear on the rocket, reducing the need for extensive refurbishment between flights. The rapid recovery and turnaround process is essential for SpaceX's long-term goal of enabling frequent, cost-effective space missions.

With each successful catch of the *Super Heavy* booster, SpaceX is proving that large-scale rocket reusability is not just possible but practical. This achievement represents a significant departure from the traditional methods of rocket recovery, where boosters are discarded after a single use or must be fished out of the ocean—a process that adds complexity, cost, and time to each mission.

SpaceX's Broader Impact on the Commercial Space Industry

Beyond its engineering marvels, *Starship* has broader implications for the commercial space industry. By lowering the cost of launching payloads into space, *Starship* opens up new opportunities for industries such as telecommunications, satellite deployment, cargo transport, and space tourism. The spacecraft's massive payload capacity—up to 100 metric tons—means it can carry multiple satellites in a single mission, lowering costs for satellite operators and enabling the deployment of large constellations such as SpaceX's *Starlink*.

In the realm of space tourism, *Starship* promises to make space travel accessible to more people than ever before. While space tourism is still in its infancy, with companies like Blue Origin and Virgin Galactic offering suborbital flights, *Starship*'s orbital and deep-space capabilities could revolutionize the industry by offering more immersive and extended space experiences. As the cost of space travel decreases and the frequency of missions increases, we could see space tourism become a mainstream industry within the next few decades.

The development of *Starship* also positions SpaceX as a key player in the global space economy. Governments, research institutions, and private companies are all looking to space as the next frontier for technological innovation, scientific discovery, and economic growth. From supporting the deployment of next-generation satellites to transporting cargo to future space stations, *Starship* is set to play a central role in this new era of space commercialization.

Scientific Exploration and Human Expansion

One of the most exciting aspects of *Starship* is its potential to advance scientific exploration. By providing a cost-effective and reliable means of launching large payloads into space, *Starship* could enable more ambitious missions to study Earth's atmosphere, other planets, and even distant galaxies. Scientific instruments and probes that were previously too expensive or large to launch could now become feasible thanks to *Starship*'s capabilities.

Moreover, *Starship* is central to humanity's long-term goal of exploring and colonizing other planets. SpaceX's role in NASA's Artemis program will be a crucial test of *Starship*'s ability to conduct manned lunar missions. If successful, these missions could lay the groundwork for the construction of a permanent lunar base, providing a platform for future research, resource extraction, and exploration of the solar system.

But it's Mars that remains the ultimate destination. Elon Musk's vision of making life multi-planetary has always been centered on establishing a self-sustaining colony on Mars. *Starship*'s design, with its reusability, large payload capacity, and ability to land on and return from other celestial bodies, is the vehicle that will make this dream a reality. While the challenges of sustaining life on Mars are immense—ranging from radiation exposure to the lack of breathable air—SpaceX's incremental progress with *Starship* brings us closer to overcoming these obstacles.

The Road Ahead: Milestones to Watch

Looking to the future, several key milestones will determine the success of *Starship* and its role in shaping the future of space exploration. The first and most immediate milestone will be the successful demonstration of orbital refueling. This process is crucial for enabling missions to Mars, as it allows *Starship* to leave Earth fully fueled and capable of long-duration spaceflight. Orbital refueling tests are expected in the near future, following the first orbital flight of *Starship* and the gathering of critical data on its performance in space.

In the coming years, SpaceX will also focus on preparing *Starship* for manned missions to the Moon as part of NASA's Artemis program. These missions will be a critical test of *Starship*'s ability to transport crew and cargo to the lunar surface and serve as a stepping stone for more ambitious missions to Mars.

Meanwhile, SpaceX continues to push the boundaries of rocket technology by iterating on *Starship*'s design and expanding the capabilities of the *Super Heavy* booster. As the number of launches increases and SpaceX gathers more data, the company will refine its systems, further reducing costs and increasing the frequency of missions.

Environmental and Regulatory Considerations

While *Starship* represents a technological triumph, SpaceX has also faced significant environmental and regulatory challenges throughout the program. The construction and testing of rockets at SpaceX's Starbase facility in Boca Chica, Texas, have raised concerns about the impact on local wildlife, wetlands, and air quality. SpaceX has worked closely with regulatory agencies like the FAA to address these concerns, conducting environmental assessments and implementing mitigation measures to protect sensitive habitats.

As the *Starship* program scales up, balancing the need for rapid innovation with environmental responsibility will remain a challenge for SpaceX. The company's focus on reusability is a positive step toward reducing the environmental footprint of space travel, but the environmental impact of frequent launches and expanding facilities will need to be carefully managed.

In addition, SpaceX will need to continue navigating the complex regulatory landscape that governs spaceflight. As the company expands its operations to new launch sites and increases the frequency of missions, it will need to work closely with regulators to ensure compliance with safety, environmental, and operational standards.

The Future of Humanity in Space

The development of *Starship* is not just about advancing technology—it's about expanding humanity's presence in the cosmos. SpaceX's vision of making life multi-planetary represents a bold and hopeful future, where humans are no longer confined to Earth but are free to explore and settle on other worlds. This vision has profound implications for the future of human civilization, from ensuring our survival in the face of potential existential threats to unlocking new resources and opportunities beyond our home planet.

As *Starship* continues to evolve, the possibilities for space travel, exploration, and human expansion will only grow. Whether it's building a permanent base on the Moon, establishing the first human colony on Mars, or exploring the outer planets, *Starship* is the vehicle that will take us there. The success of the *Starship Engine Core* in returning to base, using the chopsticks mechanism, is just the beginning of a new era in space exploration—one where the stars are no longer out of reach, but within our grasp.

A New Dawn for Space Exploration

In conclusion, SpaceX's *Starship* system represents the culmination of decades of innovation, determination, and visionary thinking. The ability of the *Starship Engine Core* to return to base, coupled with the reusability of the *Super Heavy* booster and the scalability of the entire system, has redefined what is possible in space travel. As SpaceX continues to refine and expand *Starship*, we are witnessing the dawn of a new era in which space exploration, scientific discovery, and even human colonization of other planets become achievable goals.

The journey is far from over, but with each milestone, SpaceX brings humanity one step closer to realizing its dream of becoming a multi-planetary species. As we look to the future, the possibilities are endless—and *Starship* is the vessel that will carry us there.

www.ingramcontent.com/pod-product-compliance
Lightning Source LLC
Chambersburg PA
CBHW052209220526
45471CB00004B/1880